B[e] STARS

ASTROPHYSICS AND SPACE SCIENCE LIBRARY

VOLUME 233

B[e] STARS

Proceedings of the Paris Workshop held from 9–12 June, 1997

Edited by

ANNE MARIE HUBERT
Observatoire de Paris-Meudon, France

and

CARLOS JASCHEK
Observatoire de Strasbourg, France

KLUWER ACADEMIC PUBLISHERS
DORDRECHT / BOSTON / LONDON

A C.I.P. Catalogue record for this book is available from the Library of Congress.

ISBN 978-90-481-5080-9

Published by Kluwer Academic Publishers,
P.O. Box 17, 3300 AA Dordrecht, The Netherlands.

Sold and distributed in North, Central and South America
by Kluwer Academic Publishers,
101 Philip Drive, Norwell, MA 02061, U.S.A.

In all other countries, sold and distributed
by Kluwer Academic Publishers,
P.O. Box 322, 3300 AH Dordrecht, The Netherlands.

Printed on acid-free paper

CONTENTS

PREFACE

The idea of having a meeting came to the Editors when working on several aspects of galactic Be and B[e] stars. They found that a general summary of the properties of B[e] stars was missing, so that the organization of a first meeting on these objects appeared as very useful.

B[e] stars have hydrogen line emission and forbidden [Fe II] and [O I] emission lines in their spectra; they are also characterized by a strong IR excess due to circumstellar dust.
Having a large amount of extinction in the UV and the visual they have been less frequently observed than other emission line objects. Although about one hundred galactic objects have been classified as B[e], only fifteen or so have been studied in some detail. Besides this, the evolutionary status of these objects is rather controversial, are they pre-main sequence or stars on the way to become nuclei of planetary nebulae? Other difficult problems appear when considering the relations of these stars with other similar groups, like Herbig AeBe stars, Be, luminous blue variables and the superluminous B[e] stars observed in the Magellanic Clouds.

The conference seems timely since large surveys like DENIS and 2Mass, plus new space and new instruments for the micron, millimeter and centimeter wavelength region will result in the discovery of more stars of this group.

The Scientific Organizing Committee was constituted by J. Cassinelli (Wisconsin, USA), L. Houziaux (Mons, Belgium), A.M. Hubert (Paris-Meudon, France), C. Jaschek (Strasbourg, France), H. Lamers (Utrecht, The Netherlands), J.M. Marlborough (Ontario, Canada), and A. Miroshnichenko (Pulkovo, Russia). The members of this Committee defined the scientific program and suggested invited speakers. They are to be thanked for the important contribution toward an efficient organization of the program. An efficient handling of sessions was performed by the chairpersons (J. Cassinelli, L. Houziaux, A.M. Hubert, C. Jaschek, J.M. Marlborough, A. Miroshnichenko, and R. Viotti).

The meeting was organized mainly on observational approach. It consisted of 7 sessions with overviews of general topics like definition, galactic properties, spectral and light characteristics, models, and evolutionary status of B[e] stars. Each session started with an invited paper, followed by a number of contributions, which were also shown in poster form. The meeting was closed by a summary talk by H. Lamers.

We also thank Dr. M. Combes, President of the Paris Observatory and Pr. A. Omont, Director of the Astrophysical Institute of Paris, who opened the meeting welcoming the participants.

The Local Organizing Committee was composed by D. Ballereau, D. Briot, J. Chauville, M. Floquet, P. Hammes, A.M. Hubert (Chairman), and J. Zorec. It was supported at the workshop by A. Moujtahid and M. Poinse. Logistical tasks in the form of large mailings were carried out by A. Mekkas. All the arrangements for registrations, hotels, coffee breaks and reception were handled by P. Hammes. Help for preparation of publication of these proceedings was provided by M. Floquet and V. Sabas. We thank all of them for their friendly and efficient contribution.

We like also to thank organizations who contributed financially; the presence of delegates of many countries was facilitated through a number of grants funded by the CNRS Centre National de la Recherche Scientifique, the Paris Observatory, and the CNRS Research Group 968 (Milieux Circumstellaires).

The meeting took place in the building of the Institut d'Astrophysique of the CNRS.

December 1997

A.M. Hubert and C. Jaschek

LIST OF PARTICIPANTS

BALLEREAU, Dominique	Meudon, France	ballereau@obspm.fr
BESKROVNAYA, Nina	St Petersburg, Russia	beskr@pulkovo.spb.su
BITTAR, Jamal	Toulouse, France	bittar@obs-mip.fr
BJORKMAN, Jon	Toledo, OH, USA	bjorkman@jura.astro.utoledo.edu
BOURET, Jean-Claude	Toulouse, France	bouret@obs-mip.fr
BRIOT, Danielle	Paris, France	briot@obspm.fr
CASSINELLI, Joseph	Madison, WI, USA	cassinelli@madraf.astro.wisc.edu
CHALABAEV, Almas	Grenoble, France	chalabae@gag.observ-gr.fr
CHAUVILLE, Jacques	Meudon, France	chauville@obspm.fr
CRUZADO, Alicia	La Plata, Argentina	acruzado@fcaglp.fcaglp.unlp. edu.ar
DE ARAUJO, Francisco	Rio de Janairo, Brazil	araujo@dagel.on.br
DE WINTER, Dolf	Madrid, Spain	dolf@astro2.ft.uam.es
EENENS, Philippe	Puebla, Mexico	eenens@tonali.inaoep.mx
FLOQUET, Michèle	Meudon, France	floquet@obspm.fr
FREMAT, Yves	Mons, Belgique	fremat@umh.ac.be
FRIEDJUNG, Michael	Paris, France	friedjung@iap.fr
HOUZIAUX, Leo	Mons, Belgique	leo.houziaux@umh.ac.be
HUBERT, Anne Marie	Meudon, France	anne-marie.hubert@obspm.fr
ISRAELIAN, Garik	La Laguna, Spain	gil@ll.iac.es
JASCHEK, Carlos	Strasbourg, France	
KALLEMOV, Bakhitshan	Almaty, Kazakhstan	kurt@afi.academ.alma-ata.su
KURATOV, Kenesken	Almaty, Kazakhstan	kurt@afi.academ.alma-ata.su
LAFON, Jean-Pierre	Meudon, France	lafon@mesiob.obspm.fr

LAMERS, Henny	Utrecht, Netherlands	hennyl@sron.ruu.nl
LANGER, Norbert	Potsdam, Germany	ntl@leo.astro.physik.uni- potsdam.de
LANZ, Thierry	Utrecht, Netherlands	lanz@fys.ruu.nl
LEISTER, Nelson	Sao Paulo, Brazil	leister@sismo.iapusp.usp.br
MARLBOROUGH, J. Michael	London, Canada	marlboro@titan.astro.uwo.ca
MIRABEL, Felix	Saclay, France	mirabel@ariane.saclay.cea.fr
MIROSHNICHENKO, Anatoly	St Petersburg, Russia	anat@pulkovo.spb.su
MOUJTAHID, Ali	Meudon, France	moujtahi@mesioq.obspm.fr
OUDMAIJER, René	London, UK	r.oudmaijer@ic.ac.uk
OWOCKI, Stan	Newark, DE, USA	owocki@bartol.udel.edu
POGODIN, Mikhail	St Petersburg, Russia	pogodin@pulkovo.spb.su
SADYKOVA, Sholpan	Almaty, Kazakhstan	kurt@afi.academ.alma-ata.su
SCHULTE-LADBECK, Regina	Pittsburgh, PA, USA	rsl@phyast.pitt.edu
SHEJKINA, Tatyana	Almaty, Kazakhstan	alik@itte.asdc.kz
SIMON, Guy	Paris, France	simon@obspm.fr
STEE, Philippe	Caussols, France	stee@altair.obs-azur.fr
VAKILI, Farrokh	Caussols, France	vakili@rigel.obs-azur.fr
VAN BLOKLAND, Andre	Amsterdam, Netherlands	
VIOTTI, Roberto	Frascati, Italy	uvspace@saturn.ias.fra.cnr.it
WATERS, Rens	Amsterdam, Netherlands	rensw@astro.uva.nl
YUDIN, Ruslan	St Petersburg, Russia	pulkovo.spb.su
ZICKGRAF, Franz-Josef	Strasbourg, France	zickgraf@astro.u-strasbg.fr
ZOREC, Janez	Paris, France	zorec@iap.fr

CURRENT DEFINITION OF B[e] STARS

F.-J. ZICKGRAF
Observatoire de Strasbourg
11, rue de l'Université, F-67000 Strasbourg, France

1. Introduction

About 25 years ago is was realized that an enigmatic class of emission-line stars exist, which are nowadays known as B[e] stars. A closer look to the meanwhile quite extensive literature relevant to the subject reveals, however, some confusion about which stars should be called B[e] stars. The aim of this review is to summarize from an observational point of view how these stars are presently defined and what their relation is to other classes of objects sharing similar observational properties. After a short retrospect on the history of the designation "B[e]" in Sec. 2, the defining features of B[e] stars and problems arising from the definition are discussed in Sec. 3. Similarities and differences, respectively, of galactic B[e] stars with other known object classes like classical Be stars, Herbig Ae/Be stars, symbiotic stars, and compact low-excitation proto-planetary nebulae are discussed in Sec. 4. The well-defined group of B[e] supergiants, both extragalactic and galactic, are considered in Sec. 5. Finally, conclusions are given in Sec. 6.

2. A short history of "B[e]" or "B[e] archeology"

In 1970 S.L. Geisel reported on results of an investigation of a sample of emission-line stars in a paper entitled "Infrared excesses, low-excitation emission lines, and mass loss" (Geisel 1970). Her statement that "A strong correlation has been found between stars whose continua exhibit excess radiation at infrared wavelength > 5μm and stars whose spectra contain low-excitation emission lines, especially those of Fe II and [Fe III]" in retrospect appears to mark the beginning of B[e] star research. Two years later Allen & Swings (1972) published a paper about "Infrared excesses and forbidden emission lines in early-type stars". Similar to Geisel (1970) they stated that "Low-excitation forbidden emission lines have been detected in

A. M. Hubert and C. Jaschek (eds.), B[e] Stars, 1–19.

the spectra of early-type stars found to have prominent infrared excesses." This paper was followed by a series of publications about peculiar early-type emission line stars with infrared excesses by D.A. Allen, J.P. Swings, and I.S. Glass in the early and mid seventieth. Extensive infrared surveys of galactic early-type emission-line stars, in particular by Allen (1973, 1974), and Allen & Glass (1974, 1975) revealed that in fact two distinct populations of this class of stars exist, namely emission-line stars with normal stellar infrared colours, and a group of emission-line stars with IR excesses due to the presence of hot dust. The basic observational difference of both groups is the absence or presence of this near/mid IR excess.

In 1976 the paper by Allen & Swings (1976) about "The Spectra of Peculiar Be Stars with Infrared Excesses" in a sense summarized the knowledge at that time of this class of objects. They found that these stars "... form a group spectroscopically discerned from normal Be stars and planetary nebulae... A variety of spectral features are found ...(objects) ranging from almost conventional Be stars to high density planetary nebulae. In many stars the dominant spectral features are emission lines of singly ionized iron. Hot dust ... is most often found around stars of spectral types B and early A ...".

So far, the dusty emission-line stars had been called mainly "early-type peculiar" or "Bep" stars. P.S. Conti suggested for the first time during the general discussion of the IAU Symposium "Be and shell stars" in 1976 the designation "B[e]": " ... A second class of objects would be those B-type stars which show forbidden emission lines and I would suggest that we classify these as B with a small e in brackets B[e], following the notation for forbidden lines" (Conti 1976).

At about the same time Wackerling (1970) and Ciatti et al. (1974) introduced the BQ[ ] classification for the same type of objects. More recently, de Winter et al. (1990) proposed the "Extreme Emission-Line Objects" or EELOs which also overlap in their appearance with the B[e] stars. However, these terms were not taken notice of as much as Conti's suggestion for the designation "B[e] star" which since then was generally used for the former "peculiar emission-line stars with IR excess".

The identification of B[e] type stars in the Magellanic Clouds (e.g. Zickgraf et al. 1985, 1986) added a new motive to B[e] star research, namely the investigation of B[e] supergiants. Unlike the previously studied objects they offered the immense advantage of knowing their distances and hence luminosities.

3. Definition of B[e] stars

3.1. DEFINING PROPERTIES

Based on the mentioned IR surveys Allen & Swings (1976) collected a sample of peculiar Be stars with infrared excesses which as they stated excluded young objects like Herbig Ae/Be stars, symbiotic stars (see Sec. 4) and Wolf-Rayet stars. Spectroscopically they divided the sample into three sub-groups indicating increasing excitation in the emission-line forming region:

- Group 1: few emission lines, not always forbidden lines, almost conventional Be stars
- Group 2: "most distinctive group", spectra with permitted and forbidden Fe II emission
- Group 3: additionally emission lines of higher ionization stages ($IP >$ 25eV)

Taking the properties of this sample into account and using Conti's "B[e]" designation, B[e] stars are generally defined as *early-type emission lines stars with low-excitation permitted emission lines, forbidden lines, and hot dust visible in the near and mid infrared.*

A closer look to the defining properties shows that B[e] stars, or to be more precisely, stars showing the B[e] phenomenon (see below), are characterized by the following features:

- strong Balmer emission lines, sometimes with P Cygni profiles; equivalent width $W(\mathrm{H}\alpha)$ up to $\sim 1000\text{Å}$;
- low-excitation permitted emission lines, predominantly of singly ionized metals, in particular of Fe II;
- narrow forbidden emission lines of [Fe II] and [O I];
- line widths of the low-excitation emission lines, both permitted and forbidden, typically of the order of $FWHM \approx 100\ \mathrm{km\,s^{-1}}$ or less.
- a strong near/mid infrared excess due to hot circumstellar dust, indicating dust temperatures of ~ 1000 K.

In addition, higher ionization emission lines, of e.g. [O III], or He II, may be present (see Sec. 4) but are not a defining characteristic.

For the definition of B[e] stars the IR excess is a very important property. Plotting their location in a $(J-H)-(H-K)$ or $(H-K)-(K-L)$ diagram shows evidently that they form a distinct group of objects in these two-colour diagrams. This allows to state that in connection with forbidden emission lines the presence of hot dust is ***the*** defining characteristic of B[e] stars. The $(J-H)-(H-K)$ diagram displayed in Fig. 1 (cf. also e.g. Gummersbach et al. 1995) includes all objects from Tab. 1 listed in

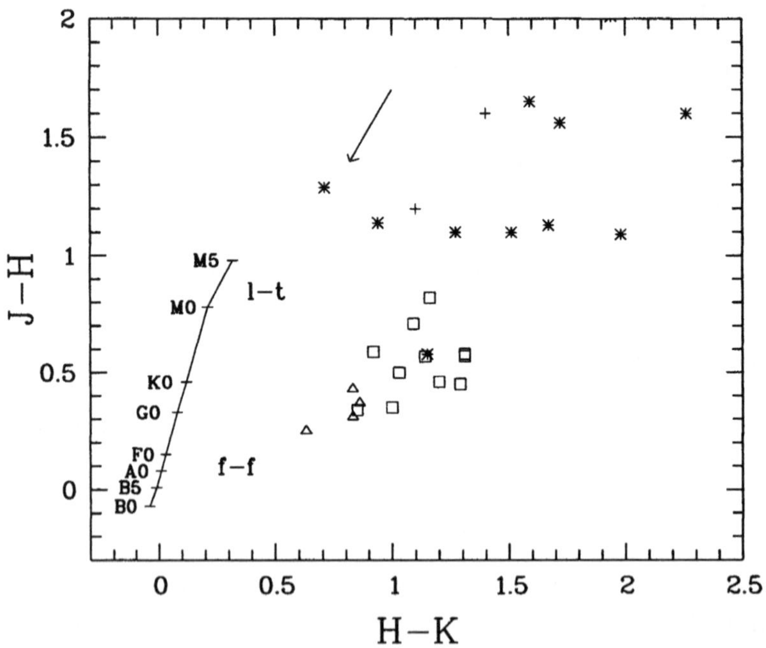

Figure 1. $(J-H)-(H-K)$ diagram for B[e] supergiants in the LMC (squares), SMC (triangles), and for several B[e] stars in the Milky Way (astrisks). The location of normal supergiants is indicated by the solid line. "f-f" denotes the approximate location of stars with IR excess due to $f-f$ emssion from stellar winds. "l-t" marks the area occupied by objects with late-type companions. The object close to the tip of the arrow which gives the reddening line for $A_V = 3$, is the binary system MWC 623 (see Sec. 4.2). Included are also two dusty symbiotic stars (crosses), RX Pup and BI Cru, which are discussed in Sec. 4.3.

Sec. 5, and furthermore the galactic stars HD 45677, MWC 17, MWC 623, HD 51585, M2-9, and Hen 1191.

3.2. PROBLEMS

This generally used definition of B[e] stars causes certain problems. As will be discussed in more detail in the following section B[e] stars obviously are **not** a homogenous group of objects. The definition as given above describes certain physical conditions in terms of excitation, density, and velocity in the circumstellar environment rather than intrinsic object properties. Because similar conditions can prevail in the surroundings of objects belonging to intrinsically different classes the definition does not uniquely describe one class of objects. It therefore seems more appropriate to speak of "stars showing the B[e] phenomenon" rather than of "B[e] stars". This should be kept in mind in the following. Because of its conciseness I will continue to used the designation "B[e] star".

The main problem occurring with B[e] stars is that in most cases their evolutionary status is unknown. This is often a consequence of unknown distances and hence luminosities. Those with known or supposedly known evolutionary status have been found to be a mixture of different object classes like the intermediate mass pre-main sequence Herbig Ae/Be stars, low-excitation compact proto-planetary nebulae which are in a post-AGB phase of low-mass stellar evolution, and B[e] supergiants in the Magellanic Clouds which represent a post-main sequence evolutionary phase of massive stars.

Furthermore, the B[e] star classification refers to single stars as being of type "B[e]". Therefore unrecognized binary systems, like symbiotic objects, can cause additional confusion.

Before going into a detailed discussion two examples, HD 45677 and MWC 300, may illustrate these problems.

HD 45677 is very often regarded as *the* proto-type B[e] star. Its spectral type is B2(III-V)e. Detailed investigations of the spectrum have been carried out by Swings (1973) and de Winter & van den Ancker (1997). Characteristic features are double-peak Fe II emission lines and single-peak forbidden [Fe II] lines. The IR excess has been ascribed to hot circumstellar dust. Polarization data, both visual and UV, indicate that the star is viewed edge-on through a dusty disk (e.g. Schulte-Ladbeck et al. 1992). Allen & Swings (1976) classified HD 45677 as proto-type of their group 2 objects, which are those showing the characteritics of B[e] stars most clearly. This suggests to regard HD 45677 as a typical B[e] star. Grady et al. (1993), however, reported to have detected mass accretion in IUE spectra. This would indicate that HD 45677 is in fact a Herbig Be star, i.e. an object in a pre-main sequence phase of evolution which should not be contained in the Allen & Swings sample. Contrary to this de Winter & van den Ancker (1997) concluded from the isolated position in the sky that HD 45677 may be a young object, but not in the sense of being a pre-main sequence object. Other authors have suggested the possibility of HD 45677 to be a "main-sequence" B[e] star (Lamers et al. 1998b).

An even more controversial case is that of MWC 300. Finkenzeller & Mundt (1984) and Thé et al. (1994)) classified MWC 300 as a "Herbig Be" star whereas Wolf & Stahl (1985) found strong evidence that it is a B1a$^+$ hypergiant, and therefore a member of the group of B[e] supergiants, which are post-main sequence objects (see Sec. 5).

These two examples show that the possible membership of B[e] stars in other classes of objects sometimes is not clear. The signatures of B[e] stars as described above can in fact be found in a variety of intrinsically very different object classes.

In the next section I will discuss in more detail similarities and differences, respectively, of objects belonging to other well-known classes with objects classified as B[e] stars.

4. Galactic B[e] stars: Similarities with other object classes

4.1. CLASSICAL Be STARS

Historically, the "classical" Be phenomenon is known since the discovery of emission lines in the spectrum of γ Cas by Secchi in 1866. For a long time the Be phenomenon has been defined by *the appearance of one emission line at one epoch in the spectrum of a star.*

Today classical Be stars are usually defined as non-supergiant B-type stars with (at some time) hydrogen lines in emission, or as O6-B9 stars of luminosity class V to III with Balmer emission lines associated with a rapidly-rotating circumstellar envelope or shell (e.g. Sterken 1996).

Be stars are more or less normal B stars with photospheric absorption features and superimposed Balmer emission lines, plus occasionally Fe II emission lines. In contrast, B[e] stars in most cases do not show normal B-type photospheric absorption line spectra. On the other hand, forbidden emission lines as always found in B[e] stars are not typical for classical Be stars. Likewise, circumstellar dust signatures are not typical for them.

Classical Be stars very often display highly variable line profiles. This is usually not found in B[e] stars which exhibit on the average only very moderate variability, if at all.

A common property of Be and B[e] stars is intrinsic linear polarization. Classical Be stars typically exhibit intrinsic polarization which can be explained by Thomson scattering in flat disk-like circumstellar geometries (e.g. Coyne & McLean 1982). Similar polarization properties were also found for B[e] stars, however with additional scattering components due to the dust (e.g. Zickgraf & Schulte-Ladbeck 1989, Schulte-Ladbeck et al. 1992) suggesting that they are also characterized by non-spherical circumstellar environments (see also Sec. 5).

Thus, Be and B[e] stars share certain characteristics, like the appearance of Balmer emission lines and the polarization properties. There are, however, also substantial differences, in particular the missing dust IR emission and the absence of pronounced forbidden lines in classical Be stars. This strongly suggests to consider both, Be and B[e] stars, as different classes of objects.

4.2. HERBIG Ae/Be STARS

The original definition of Herbig Ae/Be stars was given by Herbig (1960). According to him they are pre-main sequence objects in the mass range $2 \leq M \leq 9 - 10$ $M_\odot$ defined by following criteria:

- Spectral type A or earlier with emission lines
- location in an obscured region
- star illuminates nebulosity in immediate vicinity.

Thé et al. (1994) refined and extended this definition for their collection of a sample of Herbig stars and candidates by including F type stars, photometric properties like irregular variations, variable or high degree of linear polarization, and the presence of near or far IR excesses.

There are many similarities between Herbig Ae/Be and B[e] type stars. Finkenzeller & Mundt (1984) discussed IR photometry of Herbig stars. They exhibit strong colour excesses similar to B[e] stars. Spectroscopically, many Herbig Ae/Be stars show defining features of B[e] stars, e.g. Fe II and [O I] emission lines. The Allen & Swings B[e] star sample was collected excluding Herbig Ae/Be stars. However, many catalogs of Herbig Ae/Be stars contain objects from the Allen & Swings sample, like e.g. MWC 300 and HD 45677. This overlap of the samples could be related to the special problem of the so-called "isolated Herbig Ae/Be stars" (e.g. Grinin et al. 1989, 1991, Hillenbrand 1994). HD 45677 for example is isolated, but has been classified as Herbig star by Grady et al. (1993). Some of the possible Herbig-type B[e] stars could belong to this group. The still open question is whether they are really isolated or located in regions with young low-mass stars. Studies of the stellar environments of the B[e]/Herbig star candidates could help to clarify this problem. Hillenbrand et al. (1995) e.g. found that even apparently isolated young high-mass stars appear not to have formed alone. Rather, they seem to define centres of small aggregates of low-mass stars.

Another example is the binary system MWC 623 (Zickgraf & Stahl 1989)(see also Fig. 1). The primary component shows all signatures of a B[e] type star. The secondary, however, is a Lithium-rich K star. Hence, it seems to be rather young and therefore also the B[e] star in MWC 623 must be young. The object does, however, not share the typical Herbig star characteristics like e.g. the presence of nebulosity.

4.3. SYMBIOTIC STARS

A general feature of symbiotic stars (SS) is the appearance of "two-component" spectra which are a combination of the spectral features of a cool star and of high-excitation emission-lines, indicating the presence of a hot component. A definition of SS has been given by Allen (1979). They are

- stellar objects,
- show emission from ions with $IP > 55$ eV, i.e. He II,
- show evidence for G or later spectral type,
- if no evidence for late-type star, then lines with $IP > 100$ eV must be present, i.e. e.g. [Fe VII].

The model invoked to explain these properties is that of an interacting binary with a cool giant, usually of type M or C, a hot companion star, or compact object, and a surrounding nebula. There are some similarities with VV Cep systems which comprise a late-type supergiant plus an O or B star. Interaction leads to mass transfer which then results in the "symbiotic phenomenon". Mass accretion on the main-sequence star or white dwarf leads to the formation of an accretion disk with a hot spot. It ionizes the circumstellar material producing the observed emission lines. The spectral features of the late-type component are often only visible in infrared spectra (cf. Whitelock 1996). The emission line spectra are characterized by low-ionization lines of Fe II and [Fe II], and simultaneously high ionization lines of e.g. He II, [O III], [Ne V], [Fe VII], and [Fe X].

Symbiotic stars come in two flavours, the S- and D-type symbiotics (e.g. Webster & Allen 1975, Allen 1982):

- S-type: At wavelengths $> 1\mu$m the spectra and continuum energy distributions of these systems are generally dominated by the photosphere of the cool star. In this wavelength region they are practically indistinguishable from ordinary late-type giants. The majority of SS belongs to this subgroup.
- D-type: A fraction of about 25% of SS additionally exhibits IR emission indicating the presence of circumstellar dust. In most of the dusty symbiotic stars probably Mira variables are present. The dust appears to form in the atmospheres of the Miras. In a few symbiotics much cooler dust is present and the cool component in the system is hotter than the Miras. According to Allen (1982) in these systems the dust must form at greater distance in gas shielded from the hot component, possibly by an accretion disk.

Compared to S-type SS, D-type SS exhibit more extended, gaseous envelopes of low-enough electron density allowing the formation of a rich forbidden emission-line spectrum.

SS are irregular variables. Around maximum light the high-excitation lines disappear and an early-type shell spectrum developes, displaying metallic lines of Fe II, [Fe II], Ti II etc. in emission and absorption. Often blue-shifted absorption components of H and He I lines appear. The overall appearance is then hardly distinguishable from a B[e] star or P Cygni type star (Ciatti 1982). Therefore it is possible that some stars classified so far as B[e] stars will turn out to be symbiotic objects if they are monitored over a longer time interval. In fact, the Allen & Swings (1976) list contains several objects in their group 3 which also appear in lists of symbiotic stars, e.g. RX Pup. On the other hand, Hen 782 (BI Cru) appears as group 1 object in Allen & Swings (1976), but is now known as Mira-type symbiotic star (e.g. Harries 1996). These two stars are included in Fig. 1. Another example is the B[e] type star HD 51585 = OY Gem (Jaschek et al. 1996) which displays high-excitation lines of He II and [O III]. Because it is furthermore a photometrically variable star it could well be a symbiotic star in bright state.

4.4. LOW-EXCITATION COMPACT PROTO-PLANETARY NEBULAE

During the discussion following the talk given by Barbier & Swings on the "Be star" conference in 1982 R. Poeckert asked the question whether the B[e] stars could be a hot extension of the T Tauri stars. The interesting answer by J.P. Swings was : "My feeling is that B[e] stars are much older and are evolving towards proto-planetary nebulae..." (Barbier & Swings 1982).

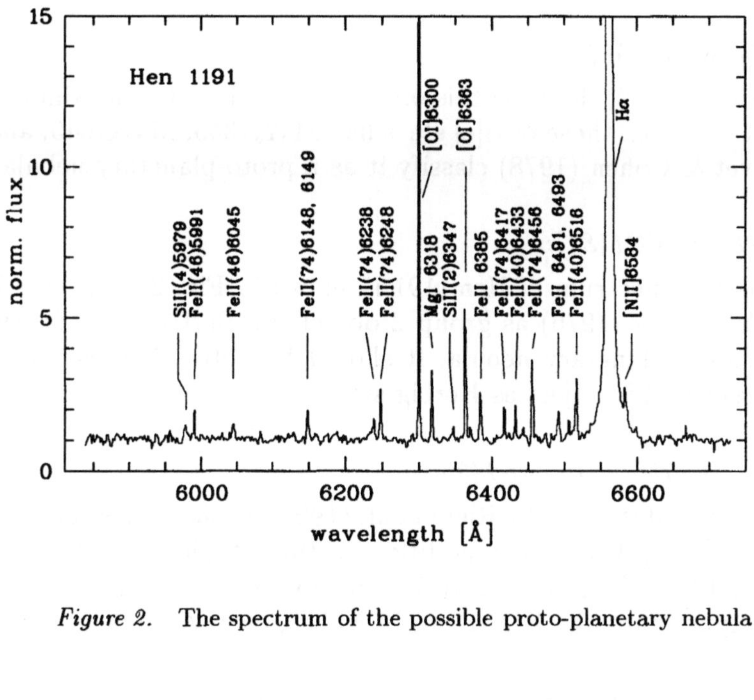

Figure 2. The spectrum of the possible proto-planetary nebula Hen 1191.

Asymptotic Giant Branch (AGB), possibly representing a post-AGB phase of evolution in the transition towards the plantary nebula stage. During the evolutionary phase near the end of the AGB the star is obscured by a dust envelope. It is created by the enhanced AGB mass loss leading to the formation of an OH/IR star. Entering the post-AGB phase the star then evolves towards the blue in the HR diagram which marks the beginning of the proto-planetary phase. While the effective temperature of the star increases further it finally becomes hot enough to ionize the circumstellar nebula. Hydrogen recombination lines and forbidden lines of metals become observable and eventually a planetary nebula will be detectable.

Various searches for objects in this transition phase of evolution made use of IRAS observations which due to the IR dust emission are well-suited to detect post-AGB objects (e.g. Manchado et al. 1989, Oudmaijer et al. 1992), in particular those with still rather cool dust envelopes. Specimen with hotter dust, however, could appear as objects showing B[e] characteristics. Ciatti et al. (1974) e.g. suggested that the BQ[ ] stars (see Sec. 2) are objects "evolving into a kind of planetary nebulae". Likewise, Parthasarathy & Pottasch (1989) suggested that BQ[ ] stars are potential candidates for these transition objects. At least a fraction of them could be hot post-AGB stars in the early stage of proto-planetary nebulae. Interestingly, 85% of the objects of group 3 of Allen & Swings (1976) are also listed in Perek & Kohoutek's (1967) "Catalogue of Galactic Planetary Nebulae" indicating that some objects in this subgroup of B[e] type stars are related to planetary nebulae. Examples are:

4.4.1. *Butterfly Nebula M2-9:*

Swings & Andrillat (1979) describe this object as showing striking similarities of its spectrum with those of B[e] stars like MWC 349, MWC 645, and MWC 819. Calvet & Cohen (1978) classify it as a proto-planetary nebula.

4.4.2. *Hen 1191 (= WR148 4):*

A section of the red spectrum of Hen 1191 is shown in Fig. 2. This star is listed in Allen& Swings (1976) as group 2 object. Le Bertre et al. (1989) classified it as proto-planetary nebula. It should be noted, however, that Thé et al. (1994) list this object as Herbig star.

4.4.3. *IRAS17423-1755 = Hen 3-1475:*

The optical spectrum described by Riera et al. (1995) displays emission lines of H I, Fe I, Fe II, [Fe II], O I, and in addition to these typical B[e] features lines of Ca II, and [Ca II]. According to the authors the spectrum appears similar to B[e] type stars like HD 51585, MWC 17, or MWC 349. Riera et al. discuss this object as a possible massive post-AGB star. They find indications for N-enhancement suggesting an evolved evolutionary stage.

5. B[e] supergiants: Extragalactic and galactic

As discussed in the previous section the galactic B[e] stars seem to be a mixture of different object classes. Many of them seem to belong to rather low-luminosity object classes. High-luminosity objects appear to be rare. One reason for this are certainly the poorly known distances. Therefore only a few galactic B[e] stars could fairly reliably be classified so far as supergiants, among them the already mentioned star MWC 300 (Wolf & Stahl 1985), and CPD−52°9243 (Swings 1981, Winkler & Wolf 1989). A section of the spectrum of CPD−52°9243 is shown in Fig. 3. In Tab. 1 five galactic B[e] supergiants are listed. It should, however, be kept in mind that for the galactic stars this list certainly is not complete.

For extragalactic B[e] stars the situation is much better. So far they have been found in the Magellanic Clouds (MCs) only. The great advantage of the MCs is that their distances and hence the luminosities of the B[e] stars and their location in the HR diagram are known. Recently, Walterbos (1997) reported on the spectroscopic discovery of B[e] candidates in M 31. However, IR photometry confirming the presence of excess radiation due to

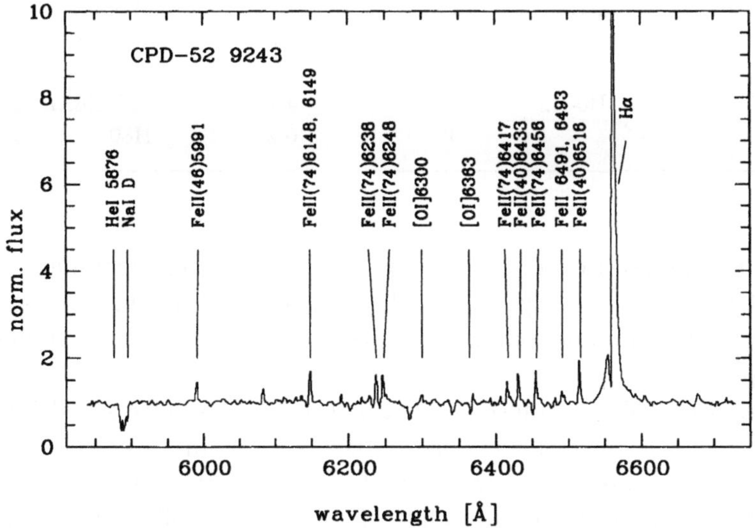

Figure 3. Section of the red spectrum of the galactic B[e] supergiant CDP−52°9243.

TABLE 1. B[e] supergiants in the LMC, SMC and the Milky Way. References are: GZW95 = Gummersbach et al. (1995), H80 = Hartmann et al. (1980), L92 = Lopes et al. (1992), McG88a,b = McGregor et al. (1988a, 1988b), S89 = Stahl et al. (1989), WS85 = Wolf & Stahl (1985), WW89 = Winkler & Wolf (1989), Z85 = Zickgraf et al. (1985), Z86 = Zickgraf etal. (1986), Z96a = Zickgraf et al. (1996a).

Name	Spec. type	T_{eff} [10^3 K]	$\log(L/L_\odot)$	reference
LMC:				
Hen S12	B0.5	23	5.3	Z86
Hen S22 = HD 34664	B0-B0.5	23-26	5.8-6.0	Z86
Hen S35	B1Iab	22	5.2	GZW95
Hen S59	B5II	14	4.0	GZW95
Hen S93	B9Ib	10	4.7	GZW95
Hen S111 = HDE 269599s	B:	–	–	McG88b, S89
Hen S134 = HD 38489	B0	26	5.9	Z86
Hen S137	B6Ib/II	13	4.2	GZW95
R66 = HDE 268835	B8	12	5.5	Z86
R82 = HDE 269217	B2-3	18.5	5.4	Z86
R126 = HD 37974	B0.5	22.5	6.1	Z85
SMC:				
Hen S18		25	5.5-5.7	Z86
Hen S23 = AV 172	B8	10-11	4.5	Z92
Hen S65 = R50	B2-3	17	5.7	Z86
R4	B0	27	5.0	Z96a
Milky Way:				
MWC300	B1Ia$^+$	20.8	5.7	WS85
CPD-52°9243	~B3Ia	17	5.5	WW89
HD 87643		16	4.9	McG88a
GG Car	B0-B2[b]	16[a]	4.7[a]	[a]McG88a, [b]L92
MWC349A		35	6.2	H80

(Stahl et al. 1989, Zickgraf et al. 1985, 1986, 1989, 1992, Gummersbach et al. 1995). A list with all presently known B[e] supergiants is given in Tab. 1. At least for the more luminous part, i.e. $L/L_\odot > 5$, the MC sample should be rather complete.

Because the interstellar extinction towards the MCs is quite small high resolution spectroscopic observations of MC B[e] supergiants were possible not only in the optical but also in the satellite UV with IUE. These observations showed the simultaneous presence of narrow ($\sim$ 100 km s^{-1}) B[e] type low-exctitation emission lines and of broad higher-excitation absorption features originating in a fast expanding ($\sim$ 1000 km s^{-1}) hot stellar wind. Zickgraf et al. (1985) suggested a two-component stellar wind model in order to explain these *hybrid* spectral characteristics. It consists of a dense cool and slowly expanding equatorial disk, and a hot and fast moving polar wind. The latter shows the characteristics of a normal line-driven OB star wind. The disk winds which have been studied by Zickgraf et al. (1996b) are about a factor of ten slower than expected from the spectral types of the stars. This could be caused by a very low effective gravity in the equatorial region, possibly due to stellar rotation.

The dust visible in the IR is formed in the disk where also molecular emission originates in some cases. Zickgraf et al. (1989) e.g. found TiO emission bands in the spectra of several MC B[e] supergiants. A section of the optical spectrum of the highly luminous B[e] supergiant Hen S134 = HD 38489 showing TiO emission is displayed in Fig. 4. McGregor et al. (1988a, b) detected CO overtone emission at 2.3μm in several stars.

Polarimetric observations showed that non-spherical circumstellar environments are common in both, galactic B[e] stars and MC B[e] supergiants. Intrinsic polarization was detected in several B[e] supergiants in the Magellanic Cloud by Magalhaes (1992). Schulte-Ladbeck et al. (1993) reported on linear polarization of Hen S18 and R50 in the SMC, and of Hen S22 in the LMC. Spectropolarimetry of Hen S22 in the LMC by Schulte-Ladbeck

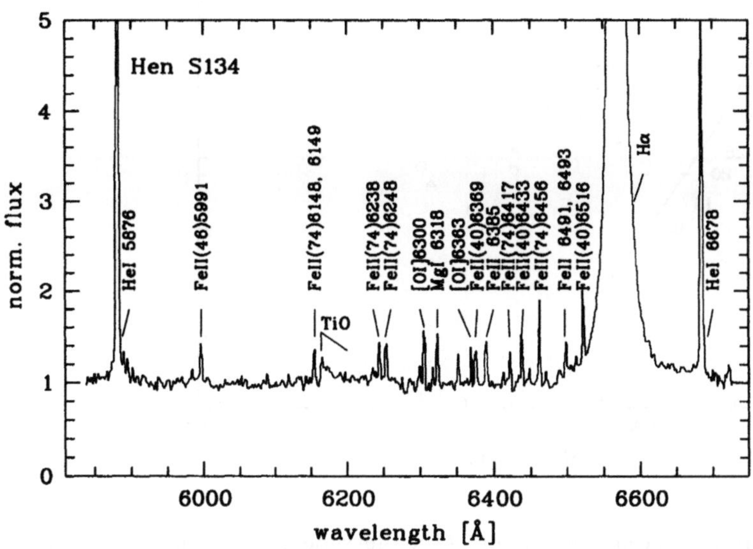

Figure 4. Section of the red spectrum of the B[e] supergiant Hen S134 in the LMC.

Linear polarization measurements of galactic B[e] stars have e.g. been performed by Coyne & Vrba (1976) who in the cases of HD 45677 found evidence for scattering from a dust ring. Barbier & Swings (1982) observed 12 B[e] stars and detected intrinsic polarization. A similar result was obtained by Zickgraf & Schulte-Ladbeck (1989) who found intrinsic polarization for a significant fraction of B[e] stars. UV spectropolarimetry of HD 45677 by Schulte-Ladbeck et al. (1992) yielded evidence for the presence of a bipolar nebula. Yudin (1996) detected variable polariztion of MWC 349, which seems to reflect changes in the ionized high-density gas. The non-variable component in MWC349 is possibly connected with the circumstellar dust. The supergiant CDP−52°9243 also exhibits intrinsic polarization (Swings 1981), in agreement with the non-spherical model suggested for the MC supergiants.

In the case of MWC 349A in the Milky Way the bipolar structure suggested by Zickgraf et al. (1985) for the MC B[e] supergiants seems to be visible. Radio observations by White & Becker (1985) revealed a bipolar stellar wind on a sub-arcsec scale with the symmetry axis perpendicular to a dusty disk found in the near IR by Leinert (1986) using speckle interferometry. 1mm radio recombination lines of maser emission with double-peaked line

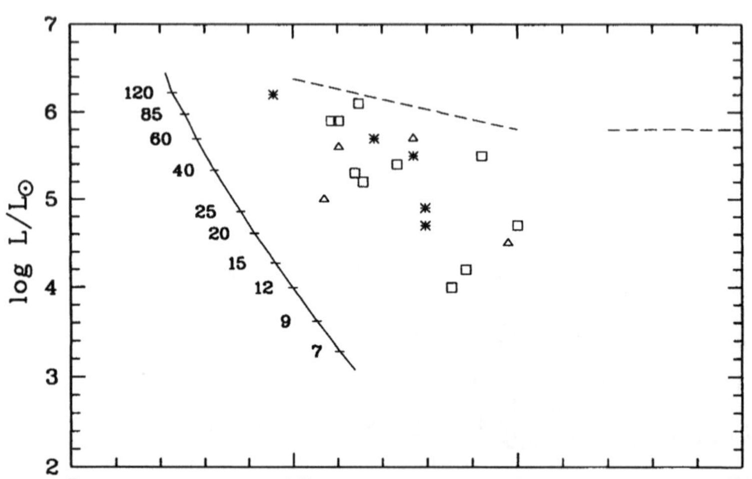

Figure 5. HR diagram showing the location of the B[e] supergiants in the LMC (squares), SMC (triangles), and the Milky Way (asterisks). The main sequence is indicated by the solid line with ZAMS mass given (Schaller et al. 1992). The dashed line represents the Humphreys-Davidson limit (Humphreys & Davidson 1994).

aration and V/R variation of the maser lines in terms of a circumstellar Keplerian disk with 22-30 AU radius which is embedded in a much larger, 200 AU radius, ionized region enclosing both, star and disk. Assuming Keplerian rotation (e.g. Hamann & Simon 1986) the double-peaked maser line profiles then lead to a central mass of $M = 34 \pm 2$ $M_\odot$ (Thum et al. 1992). MWC349A is member of a binary system in which the less massive secondary is already evolved. Therefore the more massive primary should be even more evolved. This indicates that MWC349A is very likely a massive evolved object.

Most of the B[e] stars in the MCs are extremely luminous supergiants with luminosities of $L \approx 10^5 - 10^6$ $L_\odot$. They are in a post-main sequence phase of evolution and are located in the upper left part of the HR diagram close to the Humphreys-Davidson limit (Humphreys & Davidson 1979, 1994). Recently the luminosity range could be extended downward to $L \approx 10^4 L_\odot$ (Gummersbach et al. 1995) showing that the B[e] phenomenon in the MCs occurs over a wide range of luminosity. The HR diagram showing the locations of the B[e] supergiants is displayed in Fig. 5. Direct evidence for a late evolutionary stage of B[e] supergiants was found in at least one case, R 4 in the SMC. It could be demonstrated that this B[e] star is strongly evolved having lost a large fraction of its ZAMS mass and is most likely in a post-red supergiant phase of evolution (Zickgraf et al. 1996a).

6. Conclusions

As has been shown in the previous chapters the classification of B[e] stars as used until now rather describes a certain phenomenon than a homogeneous group of objects with definite properties regarding intrinsic stellar parameters and evolutionary status. On the contrary, objects of vastly different object classes ranging from pre-main sequence to post-AGB objects and the massive post-main sequence supergiants are described by the definition of B[e] stars. Many of the galactic B[e] stars seem to be members of other already known object classes, like Herbig stars or symbiotic objects and should possibly be considered as such. Some are of unknown nature. Only the B[e] supergiants appear to represent a homogeneous group of stars which is intrinsically characterized by the B[e] phenomenon. This leaves us with the perhaps heretic question whether the B[e] supergiants are in fact the only group of B[e] type stars which really deserves to be called "B[e] stars". In any case it is clear that a revised classification scheme for B[e] stars is badly needed. Suggestions for an improved classification based on the discussions during this workshop have been made by Lamers et al. (1998a, 1998b).

References

Allen D.A., 1973, MNRAS 161, 145
Allen D.A., 1974, MNRAS 168, 1
Allen D.A., 1979, IAU Coll. 46, p. 125
Allen D.A., 1982. In: The Nature of Symbiotic Stars. Eds. M. Friedjung and R. Viotti, D. Reidel Publishing Company, Dordrecht, p. 27
Allen D.A., Glass I.S., 1974, MNRAS 167, 337
Allen D.A., Glass I.S., 1975, MNRAS 170, 579
Allen D.A., Swings J.P., 1972, ApL 10, 83
Allen D.A., Swings J.P., 1976, A&A 47, 293
Barbier R., Swings J.P., 1982. In: Be Stars. IAU Symp. 98, D. Reidel, Dordrecht, p. 103
Calvet N., Cohen M., 1978, MNRAS 182, 687
Ciatti F., D'Odorico S., Mammano A., 1974, A&A 34, 181
Ciatti F., 1982. In: The Nature of Symbiotic Stars. M. Friedjung & R. Viotti (eds.), D. Reidel, Dordrecht, p. 61
Conti P.S., 1976. In: Be and shell stars, IAU Symp. No. 70, ed. A. Slettebak, D. Reidel, Dordrecht, p.447
Coyne G.V., Vrba F.J., 1976, ApJ 207, 790
Coyne G.V., McLean I.S., 1982. In: Be Stars. IAU Symp. 98, D. Reidel, Dordrecht, p.
de Winter D., Pérez M.R., Hu J.Y., Thé P.S., 1990. In: Evolution in Astrophysics, Proc. Int. Symp., Toulouse, ESA SP-310, p. 471
de Winter D., van den Ancker M.E., 1997, A&AS 121, 275
Finkenzeller U., Mundt R. 1984, A&AS 55, 109
Geisel S.L., 1970, ApJ 161, L105
Gordon M.A., 1992, ApJ 387, 701
Grady C.A., Bjorkman K.S., Shepherd D. et al. 1993, ApJ 415, L39
Grinin V.P., Kiselev N.N., Minikulov, N.K.H., 1989, SvAL 15, 1028
Grinin V.P., Kiselev N.N., Chernova G.P., Minikulow N.Kh., Voshchinnikov N.V., 1991, Ap&SS 186, 283
Gummersbach C.A., Zickgraf F.-J., Wolf B., 1995, A&A 302, 409
Hamann F., Simon M., 1986, ApJ 311, 909
Harries T.J, 1996, A&A 315, 499
Hartmann L., Jaffe D., Huchra J.P., 1980, ApJ 239, 905
Herbig G.H., 1960, ApJS 4, 337
Hillenbrand L.A., 1994. In: The Nature and Evolutionary Status of Herbig Ae/Be Stars, eds. Pik Sin Thé, M.R. Pérez, and E.P.J. van den Heuvel, ASP Conference Ser. Vol. 62, p. 369
Hillenbrand L.A., Meyer M.R., Strom S.E., Skrutskie M.F., 1995, AJ 109, 280
Humpreys R.M., Davidson K., 1979, ApJ 232, 409
Humpreys R.M., Davidson K., 1994, PASP 106, 1025
Jaschek C., Andrillat Y., Jaschek M., 1996, A&AS 117, 281
Lamers H.J.G.L.M., Zickgraf F.-J., Houziaux L., Zorec J., de Winter D., 1998a, these proceedings
Lamers H.J.G.L.M., Zickgraf F.-J., Houziaux L., Zorec J., de Winter D., 1998b, A&A, submitted
Le Bertre T., Heydari-Malayeri M., Epchtein N., Gouiffes C., Perrier C., 1989, A&A 225, 417
Leinert C., 1986, A&A 155, L6
Lopes D.F., Daminelli Neto A., de Freitas Pacheco J.A., 1992, A&A 261, 482
Magalhaes A.M., 1992, ApJ 398, 286
Manchado A., Garcia Lario P., Pottasch S.R., 1989, Ap&SS 156, 57
Martin-Pintado J., Gaume R., Bachiller R., Johnston K., Planesas P., 1993, ApJ 418, 79
McGregor P.J., Hyland A.R., Hillier D.J., 1988a, ApJ 324, 1071
McGregor P.J., Hillier D.J., Hyland A.R., 1988b, ApJ 334, 639

Oudmaijer R.D., Van der Veen W.E.C.J., Waters L.B.F.M., Trams N.R., Waelkens C., Engelsman E., 1992, A&AS 96, 625
Perek L., Kohoutek L., 1967, Catalogue of galactic planetary nebulae, Academia publishing house of the Czechoslovak Academy of Sciences, Prague, Czechoslovakia
Parthasarathy M., Pottasch S.R., 1989, A&A 225, 521
Riera A., Garcia-Lario P., Manchado A., Pottasch S. R., Raga A.C., 1995, A&A 302, 137
Schaller G., Schaerer D., Meynet G., Maeder A., 1992, A&AS 96, 269
Schulte-Ladbeck R.L., Shepherd D.S., Nordsieck K.H. et al., 1992, ApJ 401, L105
Schulte-Ladbeck R.L., Clayton G.C, 1993, AJ 106, 790
Schulte-Ladbeck R.L., Clayton G.C., Leitherer C., Drissen L., Robert C., Nota A., Parker J., 1993, SSRv 66, 193
Stahl O., Smolinski J., Wolf B., Zickgraf F.-J., 1989. In: Physics of Luminous Blue Variables, eds. K. Davidson, A.F.J. Moffat, and H.J.G.L.M. Lamers, Kluwer Academic Publishers, Dordrecht, p. 295
Sterken, C., 1996. In: Light curves of variable stars, eds. C. Sterken and C. Jaschek, Cambridge University Press
Swings J.P., 1973, A&A 26, 443
Swings J.P., 1981, A&A 98, 112
Swings J.P., Andrillat, 1979, A&A 85, 74
Thé, P.S., de Winter, D., Pérez, M.R. 1994, A&AS 104, 315
Thum C., Martin-Pintado J., Bachiller R., 1992, A&A 256, 507
Wackerling L.R., 1970, Mem. R. astr. Soc. 73, 153
Walterbos R.A.M., 1997. In: Luminous Blue Variables: Massive Stars in Transition; eds. A. Nota and H.J.G.L.M. Lamers, ASP Conference Ser. Vol. 120, p. 182
Webster B.L., Allen, D.A., 1975, MNRAS 171, 171
White R.L., Becker R.H., 1985, ApJ 297, 677
Whitelock , 1996. In: Light curves of variable stars, eds. C. Sterken and C. Jaschek, Cambridge University Press
Winkler, H., Wolf B., 1989, A&A 219, 151
Wolf B., Stahl O., 1985, A&A 148, 412
Yudin R.V., 1996, A&A 312, 234
Zickgraf F.-J., Schulte-Ladbeck R.E., 1989, A&A 214, 274
Zickgraf F.-J., Stahl O., 1989, A&A 223, 165
Zickgraf F.-J., Wolf B., Stahl O., Leitherer C., Klare G., 1985, A&A 143, 421
Zickgraf F.-J., Wolf B., Stahl O., Leitherer C., Appenzeller I., 1986, A&A 163, 119
Zickgraf F.-J., Wolf B., Stahl O., Humphreys R.M., 1989, A&A 220, 206
Zickgraf F.-J., Stahl O., Wolf B., 1992, A&A 260, 205
Zickgraf F.-J., Kovacs J., Wolf B., Stahl O., Appenzeller I., Kaufer A., 1996a, A&A, 309, 505
Zickgraf F.-J., Humphreys R.M., Lamers H.J.G.L.M., Smolinski J., Wolf B., Stahl O., 1996b, A&A 315, 510

Discussion

Owocki: I noticed that you said nothing about the rotation properties of those B[e] stars. I realize that information on this is limited because of obscuration by circumstellar material but could you please comment on what information is known?

Zickgraf: For Hen S 93 the rotation velocity could be determined (cf. Gummersbach et al. 1995). But this star is most likely viewed pole-on. In the other cases of B[e] supergiants with photospheric lines a reliable determi-

nation of the rotational velocity is not possible.
Langer: Has there been any attempt to obtain chemical abundances from B[e] supergiant spectra?
Zickgraf: For the B[e] component in R 4/SMC we obtained estimates of the CO abundances. It appears that is overabundant and C is obviously strongly underabundant.
Friedjung: How do you determine the spectral type of a B[e] supergiant? It is easy to find the height of a star in the HR diagram, but not the horizontal position. Without a model you have a problem. In old work I was involved in, there were indications of an optically thick disk contributing to the optical continuum in some cases.
Zickgraf: In some cases photospheric absorption features are visible which allow classification (R 50/SMC, R 82/LMC, Hen S 93/LMC). But most B[e] supergiants were classified based upon the UV to optical spectral energy distribution, usually using LTE model atmospheres. In the cases of Hen S 93 and R 4 detailed modeling of the photospheric absorption spectrum was possible (see Gummersbach et al 1995 and Zickgraf et al 1996).
Marlborough: Is there any evidence that the stars observed by Gummersbach et al have undergone massive mass loss?
Zickgraf: no there is no indication for that.
Lamers: The B[e] supergiants in the HR diagram of the MC seem to fall into two groups, separated at about log $L \approx 4.5$ Do you think that this indicate two distinct groups? Are they different in any other characteristics?
Zickgraf: The apparent gap in the HR diagram could be a consequence of the small number statistics. However, because we surveyed about one hundred emission line stars in the MCs in order to search for B[e] stars, the absence of such stars in the gap could in fact be real. The lower luminosity stars are not different in a systematic way, with the exception that two of them show more pronounced photospheric absorption features.
Cassinelli: I do not understand the evolutionary state of the two stars in the binary system R 4. Is the A star a post red supergiant star? or is it the B[e] star?
Zickgraf: It is the B[e] star which is supposed to be in the post red supergiant phase. The binary component shows evidence for strong mass loss and for nitrogen enrichment.
Houziaux: Have you been able to observe the Balmer discontinuity in the spectra of B[e] supergiants either galactic or in the MCs? The discontinuity might be an indication of the photospheric spectral type. What about the reddening of luminous galactic B[e] stars versus similar objects in the Clouds?
Zickgraf: R 50 for example shows a Balmer jump indicative of a late B type supergiant. The photospheric spectrum and the UV continuum how-

ever lead to a B2-3 spectral type- The galactic B[e] stars are on the average strongly reddened.

Langer: As the A type component of R 4 is clearly in the post main sequence stage, it seems impossible (at first glance) that the much more luminous B[e] component has not yet reached the supernova stage. Strong mass loss of the B[e] component does not help here.

Araujo: Two points I suppose to be related. How do we compare UV observations of luminous extragalactic B[e]'s and those of galactic B[e]'s and do you suggest the two component wind model only for the extragalactic superluminous or for both groups?

Zickgraf: Except HD 87643 no other B[e] supergiant was observed in the UV, at least to my knowledge. So I cannot say much about galactic B[e] supergiants. But of course the two components wind model should apply also to the galactic objects.

SEARCH FOR B[e] STARS IN YOUNG OPEN CLUSTERS

D. DE WINTER
Dpto. Física Teórica, C–XI, Universidad Autónoma de Madrid, Cantoblanco, 28049 Madrid, Spain

M.E. VAN DEN ANCKER
Astron. Inst. "Anton Pannekoek", University of Amsterdam, Kruislaan 403, 1098 SJ Amsterdam, The Netherlands

AND

M.R. PÉREZ
Space Applications Corp., Suite 250, 9315 Largo Drive West, Largo, MD 20774, USA

Abstract. The properties and the stellar content of five young open clusters have been investigated to identify new B[e] stars. After a detailed analysis we conclude that none of the ~ 260 relatively young early type stars can be identified as such. We propose that either the B[e] characteristic is not present in young stars or that due to the close proximity of other hot stars the circumstellar material can not hold sufficient amounts of material to show this spectroscopic property. After investigating the direct environment of well known galactic B[e] stars we find support for our hypothesis that the B[e] phenomenon is mostly seen in rather isolated sources. As many massive stars are located in clusters it explains the paucity of this spectral characteristic.

1. Introduction

The stellar population of five well-known galactic young open clusters has been studied: NGC 2244, 2264, 6383, 6530, 6611. The first goal was to find Herbig Ae/Be (HAeBe) type stars among the early type members of these young clusters. However, only relatively few were detected (~ 21).

For the majority of the bright visible stars (more than 250 objects) their observable properties (optical to near-IR photometry and low resolution

A. M. Hubert and C. Jaschek (eds.), B[e] Stars, 21–26.

spectroscopy), as well as other astrophysical parameters were studied in some detail. As the sample is obviously biased toward O- and B-type stars, it is also perfectly suited for the detection and identification of B[e] stars among these very young objects of unknown type.

The results of this study will be shown and discussed by presenting the characteristics of individual objects, as well as the general cluster properties. Finally, the results will be discussed in relation to known B[e] stars.

2. Description of the sample

An advantages of the use of young open clusters in these kind of studies is that no confusion exists about the distance and age of the cluster-members. This has been a major confusion for the classification of known B[e] stars, see several papers in these proceedings. Most of cluster-members are also bright enough for a wide range of observational work. Besides these advantages, there is a difficulty to discriminate between real cluster-members and field stars. For each of the clusters these problems are widely discussed and sample selections are carefully made in the studies mentioned in Table 1.

In general the selection of B[e] stars is made on the obvious spectroscopic properties, being spectral type B and the presence of emission lines of [Fe II] and [O I], and an indication of a near-infrared (NIR) or far-IR excess.

3. The observational results

We have obtained 4000–8000Å low resolution spectroscopy (LRS) and $UBVRI-JHK(LM)$ photometry of probable cluster members. IRAS data for cluster-members can not be derived, because of the low spatial resolution of the instrument and the strong background radiation. Spectral corrections for emission lines of bright nebulae in these clusters is not always easy as the emission can be intracluster rather than intrinsic.

The final spectra were studied and the spectral energy distributions (SEDs) were made and investigated. The observations and all results can be found in the original studies, see Table 1. In this table we have also summarized the properties for each cluster-sample and of each of the clusters its properties as derived in these studies.

The results as presented in Table 1 show that in none of the clusters, all very young, not even a single B[e] star was detected. Why?

4. The circumstellar environment of B[e] stars

We have seen that no B[e] stars were found in large samples of young early type stars in several very young ($\ll 10^7$ yrs) clusters. It is not an exceptional property of one of these clusters. We propose, therefore, that B[e] stars are

TABLE 1. Properties of early type stars in several YCOs. The abbreviations in the heading refer to the original studies: dW = de Winter et al. (1997); Hil = Hillenbrand et al. (1993); MP = Pérez et al. (1987, 1991); PT2 = Thé et al. (1985); PT2 = Thé et al. (1990); vdA = van den Ancker et al. (1997). For some cluster properties estimates are given with the error between parentheses.

NGC: - - - - > Characteristic	2244 MP	2264 MP	6383 PT1	6530 vdA	6611 dW+PT2	6611 Hil	Total
# Objects studied	104	80	14	132	87	332	749
LR spectra	21	22	14	80	57	67	261
Strong emission	1	1	1	7	18	20	48
Weak/possible emission	0	0	0	12	9	5	26
NIR data	21	22	14	26	49	28	160
NIR excess	1	9	5	9	12	7	43
HAeBes candidates	2	6	1	5	4	3	21
O & B stars with LRS:	21	22	5	54	37	57	196
Emission	1	1	0	13	22	19	56
NIR data	21	22	5	30	19	27	124
NIR excess	1	9	1	9	8	7	35
NIR excess & emission	0	1	0	3	6	3	13
B[e] Stars	0	0	0	0	0	0	0
Cluster properties:							
d [kpc]	1.67(1)	0.95(7)	1.5(2)	1.8(2)	2.6(3)	2.1(1)	-
interstellar $E(B-V)$	0.48(2)	0.06(1)	0.31(2)	0.30(2)	0.50(2)	0.4	-
t_{max} [Myr]	< 4	< 5	< 6	< 4	< 6	< 6	-
t_{ave} [Myr]			1.7(4)*	2(1)**	1–2	2(1)	-
t_{min} [Myr]			< 0.3	0	0	0	-

Notes:
*) Taken from Luiken-Miller (1982). **) Taken from Sagar & Joshi (1978).

not extremely young ($\gg 10^7$ yrs), or that their CS material (disk?) survival time is $\ll 10^7$ yrs.

The latter is supported by the detection of Hester et al. (1996) that dusty regions in NGC 6611 suffer from erosion due to the intense UV radiation fields of nearby O & B stars. This might work also as disk erosion of young stars in such clusters (Pérez et al. 1998). In fact their calculations show that in these crowded regions the disk survival time is less than the pre-main sequence time of a few Myrs for the studied stars, which explains the lack of detected HAeBe stars, see Table 1.

Note that most well known HAeBes are found in relatively isolated positions. What about the known B[e] stars? We have studied this and found 41 early type (ET) stars (O, B and A) within a circle of 10′ around 28 known galactic B[e] stars (de Winter & Pérez, these proceedings). However 19 of these nearby ET stars are around only 4 B[e] objects, these are: CD−42°11721, HD 87643, He 2–14 and MWC 17. As 24 B[e] stars have only 22 ET neighbours within 10′ we conclude that the known galactic B[e] stars are isolated objects, a suggestion already proposed by Zickgraf & Schulte-Ladbeck (1989) for a few B[e] stars.

5. Conclusions

From the non-detection of B[e] stars among more than 250 well studied early type stars in 5 young open clusters and the investigation of the surroundings of 28 known galactic B[e] stars we summarize the main possibilities for the nature of B[e] stars:

- **B[e] stars are young and isolated.** In this case the disk material is young and the B[e] star will be clearly located in or near a SF region, possible examples are R Mon and V380 Ori (de Winter & Pérez, these proceedings).
- **B[e] stars are evolved objects with "young disks".** The "disk" can be relatively young when it originates from mass loss. The star might be isolated or not. Examples of such objects are massive stars such as LBVs and B[e] supergiants in the MCs, and stars with recent mass-loss like post-AGB stars.
- **B[e] stars are evolved objects with "old disks".** The disk survived a significant time of the stellar evolution for some reason. The B[e] star will be isolated and does not belong to the previously mentioned groups. Examples could be HD 45677 and HD 50138.

The final conclusion is that, as for HAeBes, the environments of B[e] stars might be very important to understand the presence and the rarity of the B[e] phenomenon, which is not detected in young open clusters.

Acknowledgements

D. de Winter is supported in part by Spanish grant DGICYT PB94-0165. M. van den Ancker acknowledges financial support through an NWO *Pionier* grant to L.B.F.M. Waters.

References

van den Ancker, M.E., Thé, P.S., Feinstein, A., Vázquez, R.A., de Winter, D., Pérez, M.R. 1997, A&AS 123, 63

Hester, J.J., Scowen, P.A., Sankrit, R., Lauer, T.R., Ajhar, E.A., Baum, W.A., Code, A., Currie, D.G., Danielson, G.E., Ewald, S.P., Faber, S.M., Grillmair, C.J., Groth, E.J., Holtzman, J.A., Hunter, D.A., Kristian, J., Light, R.M., Lynds, C.R., Monet, D.G., O'Neil, E.J., Shaya, E.J., Seidelmann, K.P., Westphal, J.A. 1996, AJ 111, 2349
Hillenbrand, L.A., Massey, P., Strom, S.E., Merrill, K.M. 1993, AJ 106, 1906
Luiken-Miller, M. 1982, Ph.D. thesis, University of Waterloo, Canada
Pérez, M.R. 1991, Rev. Mex. Astron. Astrof. 22, 99
Pérez, M.R., de Winter, D. , van den Ancker, M.E., Grady, C.A., Thé, P.S. 1998, A&A, to be submitted
Pérez, M.R., Thé, P.S., Westerlund, B.E. 1987, PASP 99, 1050
Sagar, R., Joshi, U.C. 1978, MNRAS 184, 467
Thé, P.S., Hageman, T., Westerlund, B.E., Tjin A Djie, H.R.E. 1985, A&A 151, 391
Thé, P.S., de Winter, D., Feinstein, A., Westerlund, B.E. 1990, A&AS 82, 319
de Winter, D., Koulis, C., Thé, P.S., van den Ancker, M.E., Pérez, M.R. Bibo, E.A. 1997, A&AS 121, 223
Zickgraf, F.-J., Schulte-Ladbeck, R.E. 1989, A&A 214, 274

Discussion

Zickgraf: Could the non-detection of B[e] stars be a consequence of number statistics, with other words, can you estimate how many massive stars with, say 15 or more $M_\odot$, you have in your sample?

de Winter: There are many massive stars in this sample, probably about 25%, with some extremes of $\sim$50 $M_\odot$. The cluster members are very young, less than 10 Myrs, which means that most massive stars in a HRD occur close to the main-sequence and not yet evolved in the region where one find the supergiant B[e]s, like in the MCs, indeed. So in this manner it is biased, biased to very young stars, as the title suggests.

Israelian: How is it possible to separate spectroscopically from a pure emission line whether we have an accretion or an excretion disk?

de Winter: If the profiles are not resolved one cannot. Like in this study, we have only low resolution spectroscopy. It is evident that high resolution spectra is needed to resolve indications of accretion, such as inversed P Cygni profiles, such as has been detected for several Herbig Ae/Be stars in Hα.

Lamers: You find that the dusty-disk of HAeBe stars can disappear due to erosion by hot stars in the same cluster. But why don't you see forbidden emission lines? Is the presence of dust a necessary condition for the formation of [e] lines?

de Winter: This seems to be proven observationally. All known B[e] stars or candidates show to have a near- or far-infrared excess. Of course this could be a bias by selection criteria. But note that the Allen & Swings samples were selected pure on spectroscopic properties. For some B[e] stars continuum fluorescence explains the occurrence of [Fe II] emission lines. Very strong Fe II absorption lines in the UV (IUE spectra) are visible for these stars. This requires high density gas which might be continuously created by evaporation of the circumstellar dust. But I'm not sure if B[e] stars without dust are present somewhere, or that other models could explain their Fe II/[Fe II] emission line spectrum.

Cassinelli: You suggest as one option that B[e] stars are evolved with "old" disks. Could you explain what you mean by that? Could the disk arise from a previous Red Supergiant phase of the star?

de Winter: Yes, but I would classify this as recently created circumstellar material. If we do not discuss the morphology, I classified this as a young "disk". I refer to an old "disk", for material that has survived a significant time around the object, in other words, cases in which we can not (yet) indicate a preceding recent evolutionary phase in which this circumstellar material was formed. These objects can not be classified within one of the known B[e] stellar groups.

DISTANCES, KINEMATICS AND DISTRIBUTION OF B[e] STARS IN OUR GALAXY

J. ZOREC
Institut d'Astrophysique-CNRS, *98bis, bd. Arago, F-75014 Paris*

1. Introduction

A group of B and late O type stars with hydrogen emission lines also show forbidden emission lines in their visual spectral range. The presence of forbidden lines is correlated in most cases with a strong IR flux excess (Allen & Swings 1976). Although they all have the generic name of "B[e] stars", a large variety of objects may in principle be comprised in this group. A quick look at the existing lists of known B[e] stars (Jaschek & Egret 1982) reveals that the "B[e] phenomenon" concerns objects that may also show properties of "classic" Be stars (Jaschek et al. 1981), as well as those of supergiants, planetary nebulae, symbiotic and young stellar objects either in the early main sequence or in a pre-main sequence evolutionary stage. This may imply that the circumstellar envelopes giving rise to the line and continuum emissions were formed by different mechanisms and/or that the "B[e] phenomenon" concerns a different type of objects. Qualitative information on the underlying stars showing the "B[e] phenomenon" may in part be obtained from knowledge of their galactic distribution and kinematics, whose reliability depends on the quality of their distance determination. However, quantitative interpretations of their nature can sometimes be done only if good distances are known. Distances of these stars are either highly uncertain or unknown in most cases. Thus, the main concern of this paper is distance determination, and in some relation to this problem, the kinematics and galactic distribution of stars showing the B[e] phenomenon.

2. Distance determination

Apart from the few examples of B[e] stars that have parallaxes measured by the HIPPARCOS satellite, the distances reckoned for the remaining galac-

A. M. Hubert and C. Jaschek (eds.), B[e] Stars, 27–45.

tic objects of this class are highly uncertain. Another group of stars of this class with known distances is that of B[e] supergiants of Magellanic Clouds (Zickgraf 1993). The presence of dusty envelopes, strong interstellar absorption, crowding of photospheric absorption lines and continua by strong line and continuum emissions seems to make a spectroscopic estimate of distances almost impossible. So, rather than referring to reliable, well-defined methods, in most cases we need to use indirect ways and rough statistical relations to determine their distances.

2.1. INDIRECT ESTIMATION OF DISTANCE

The indirect ways most frequently used to infer distances of stars presenting the B[e] phenomenon are: (a) membership of stellar systems; (b) membership of stellar Populations; (c) stellar kinematics; (d) statistical relations between distance, colour excess and interstellar line strengths; (e) distribution with distance of interstellar (ISM) absorption.

(a) *Membership of stellar systems.* Up to now, there is only one reliable example of a B[e] star belonging to a stellar cluster: star No 9 of a compact, heavily reddened cluster at $D \simeq 5$ kpc in Ara discovered by Westerlund (1987). For a long time, MWC 349 was thought to be in the Cygnus OB 2 association at $D \simeq 2.1$ kpc (Reddish 1967), but this has since been shown to be erroneous (Cohen et al. 1985) as the star was proved to be a binary system whose distance, deduced from the B0IIIe companion, is: $D \simeq 0.97$ kpc. Oudmaijer (1998) calculated the spectroscopic distance of MWC 297 and showed that the star is associated with a molecular cloud. From their apparent position in the sky, a small number of B[e] objects may perhaps belong to R-associations. They are given in Table 1. Some B[e] stars, those listed in Table 2, seem to be associated with H II regions, but distances are not known for all of them.

(b) *Membership of stellar Populations.* The orbital movements of stars in an axisymmetric galactic potential may suffer from accelerations imposed mostly by chance encounters with other stars and massive interstellar clouds (Mihalas & Binney 1982). The observed increase of the standard deviation of peculiar velocities with stellar age is also followed by an increase of the mean absolute distance $\overline{|z|}$ from the galactic plane. If the stellar group to which the star might belong is somehow identified, using tables of $\overline{|z|}$ corresponding to different types of stars in several stellar Populations (Scheffler 1982) and the star galactic coordinates (l,b), the distance is inferred from $D \sim \overline{|z|}\, csc\, b$. However, given uncertainties leading to the choice of the appropriate $\overline{|z|}$ and the knowledge that the star may actually have any value of z, from 0 to sometimes $\overline{|z|}$, the above relation leads to quite uncertain values of D.

TABLE 1. B[e] stars possibly associated with R-associations

Star	$e(b-v)$	D(pc)	R-association	l°	b°
MWC 342	0.58	1000	Cyg R1	77	+2
Hen 3-230	0.61	870	Vela R2	265	+2
Hen 3-91	0.64	2190	Cen R1	305	0

Herbst (1975a,b), Vega et al. (1986)

(c) *Stellar kinematics.* The stellar kinematic distance can be inferred from Oort's relations of galactic differential rotation. However, proper motions are known for quite a few B[e] stars and radial velocities are highly uncertain. In most cases no absorption line is seen in B[e] stars, so the radial velocities are obtained from more or less variable emission lines. Moreover, only mean values of kinematic distances of stellar groups for which the mean value of peculiar velocities is zero have some meaning.

(d) *Statistical relations between distance, colour excess and line strengths.* The intensity of interstellar Ca II K and Na I D1,D2 lines is correlated with distance (Hobbs 1974, Scheffler 1982, Welty et al. 1994). In the same way, relations were established between equivalent widths and/or central intensities of diffuse interstellar molecular bands (Herbig 1975, Snow et al. 1977, Sneden et al. 1978, Scheffler 1982, Herbig and Leka 1991). These mean relations were more or less currently used to infer distances of B[e] stars. However, no single-valued relations between interstellar diffuse bands and colour reddening appear to exist, as the carriers producing the extinction would be not the same as those giving rise to the diffuse interstellar bands. There may also be some doubt as to whether the interstellar absoption lines are formed in the same environment where dust exists. The density distribution of the interstellar medium (ISM) is very irregular (Neckel & Klare 1980) and its chemical composition is not homogeneous (Herbig 1975). The mean relations between equivalent widths, E(B-V) and distances may then be not the same for all directions. So, these relations are better defined if they are determined for a selected direction (Herbig 1975).

(e) *ISM absorption distribution.* Using the stars projected in the sky in the near surroundings of the studied B[e] object, the distance dependence of E(B-V) can be obtained towards the stellar direction. Hence, once the $E(B-V)_{ism}$ of the studied B[e] star is inferred, its distance can easily be estimated from the obtained relation. Apart from the uncertainty introduced by the scatter of points in the [E(B-V),D] diagram, which is mainly due to non-homogeneous density and chemical composition distribution of interstellar clouds, there is also an intrinsic uncertainty affecting the E(B-V) obtained for a studied B[e] star. This last arises as the true ISM component of the observed E(B-V) is not easily disentangled from the perturbing reddening components due to emissions and absorptions produced by the gaseous and dusty stellar circumstellar envelopes. However, most of the distances of B[e] stars known today were estimated using the [E(B-V),D]-method.

TABLE 2. B[e] stars associated with H II regions

Star	H II region	l°	b°
MWC 56	S 193	136.1	+2.1
MWC 297	S 62	26.8	+3.54
MWC 623	S 98	68.2	+1.02
MWC 137	S 266	195.7	-0.1
MWC 349	S 109	79.5	+0.2
MWC 342	S 108	78.2	+1.8

Blitz et al. 1982; Sharpless 1959

2.2. SEARCH FOR A CONSISTENT DISTANCE DETERMINATION

The above indirect methods do not give satisfying distance estimates even for "normal" stars. Most known stars with the B[e] phenomenon are very distant, so that direct parallaxes were measured by the satellite HIPPARCOS for only a few of them. It is well known that spectroscopic distances of normal stars have a probable error of about 20 %. Seeing all the unknowns which still obscure the B[e] phenomenon, it seems almost hopeless to seek a spectroscopic-like method of distance determination which can be used for these stars. Nevertheless, it is worth putting together the known characteristics of these objects and trying to obtain, as far as possible, a consistent framework for distance determination.

2.2.1. *Photometric properties of B[e] stars*

In the past, long-term photometric surveys of B[e] stars were done only for a few examples (Arkhipova 1989, Arkhipova & Ikonnikova 1992, de Winter et al. 1996). Apart from these rare cases, the most complete long-term and homogeneous photometric data for B[e] stars were recently published by Bergner et al. (1995). Almost always, the existing photometric surveys show that over long periods of time the V magnitude and the (B-V) colour index are almost unchanged. The mean standard deviation of their variations are respectively: $\overline{\sigma_v} \simeq 0.08$ mag and $\overline{\sigma_{b-v}} \simeq 0.14$ mag. There is an exception, HD 45677, which from about 1953 to 1982 showed a brightness decrease of about 1.6 mag and has still not completely recovered its former quasi-static apparent magnitude (de Winter & van den Ancker 1997). Superimposed on this long-term photometric variation, the star also has variations of shorter time scales (days, hours) with amplitudes $\Delta v \sim 0.2$ mag and $\Delta(b-v) \sim 0.05$ mag. In general, the existing photometric data for B[e] stars do not seem to show long-term variations resembling those of "classic" Be stars. During definite emission phases and during "shell" phases, classic Be stars are characterized by different $\Delta v/\Delta(b-v)$ and $\Delta(b-v)/\Delta(u-b)$ slopes (Moujtahid et al. 1997). Such differentiated photometric behaviour, if detected in B[e] stars, would allow us to determine their approximate photometric spectral type (Moujtahid et al. 1998a). Finally, let us note that in the visible spectral range of B[e] stars there are strong emission lines which sometimes markedly affect the photometric indices. Hence, variation of these lines contributes to spurious photometric variations. Spectral energy distributions like those obtained by Swings (1981), that would be suitable for a larger number of B[e] stars, help to correct the effect of lines on V and (B-V).

2.2.2. *Pogson's relation for B[e] stars*

Strong Balmer and Paschen emission lines in B[e] stars (Jaschek 1998) reveal that a gaseous component of the circumstellar envelope (GCE) must also contribute with (b-f) and hot (f-f) emission/absorptions to the observed continuum energy distribution. In "classic" Be stars, the flux excess in the V magnitude, Δv_g, is supposed to be due to the GCE only. It can be estimated measuring the second Balmer discontinuity (Moujtahid et al. 1997) or the $H\gamma$ line emission component (Ballereau et al. 1995). The magnitude Δv_g and colour $\Delta(b-v)_g$ excesses are correlated with the effective temperature and they are also a function of the aspect angle i (Zorec & Briot 1991, 1997, Moujtahid et al. 1998c). The Δv_g and $\Delta(b-v)_g$ excesses averaged over the aspect angle are:

$$\Delta v_g \simeq [1.67 + 3.49\,(T_{eff}/10^4)]\,\Delta\mathcal{D} \quad mag \qquad (1)$$

$$\Delta(b-v)_g \simeq -[0.55 + 1.18\,(T_{eff}/10^4)]\,\Delta\mathcal{D} \quad mag \qquad (2)$$

where $\Delta\mathcal{D} \leq 0$ in *dex* represents the second Balmer discontinuity. Both relations above are for 10^4 K $\leq T_{eff} \leq 3.5\,10^4$ K. $\Delta\mathcal{D}$ is also temperature-dependent and typical values of $\Delta\mathcal{D}$ in B[e] are: $0 \leq \Delta\mathcal{D} \leq 0.7$ dex. We can also write (Ballereau et al. 1995):

$$\Delta v_g \simeq -2.5\,\log[1.57 + 1.7\,W_{H\gamma}] \quad mag \qquad (3)$$

where $W_{H\gamma}$ is the equivalent width of the $H\gamma$ line emission component and the relation holds for $W_{H\gamma} \leq 3.5$ Å.

Thus, to describe the observed V magnitude of B[e] stars we may think of an underlying normal stellar energy distribution, affected by an emission excess due to the GCE. The resulting energy distribution is still modified by the dusty circumstellar envelope (DCE) and finally, it is more or less strongly absorbed by the interstellar medium (ISM). Pogson's relation for the magnitude V is then:

$$v = (M_V + \Delta v_g) - 5 + 5 \log D + A_v^T \tag{4}$$

where M_v is the visual absolute magnitude, which, as far as we have no better representation, is considered to be that of a "normal", non-rotating star: $M_v = f(T_{eff}$, luminosity class). The same approximation is also used for the intrinsic colour index $(B\text{-}V)_* = f(T_{eff}$, luminosity class). The Δv magnitude excess is considered as a modification of the visible absolute magnitude, so that the total absorption A_v^T is left with two components:

$$A_v^T = R_{ism}\, e(b-v)_{ism} + R_{dce}\, e(b-v)_{dce} \tag{5}$$

where the R's are the ratios of total A_v to selective $E(B\text{-}V)$ absorptions. In general, R_{ISM} is a function of galactic coordinates, but for most cases we can assume a mean ISM absorption law with $R_{ism} = 3.1$ (Mathis 1990). The $E(B\text{-}V)_{ism}$ colour excess is a function of the distance D and of the observed direction (l,b): $E(B\text{-}V)_{ism} = f(D|l,b)$. This relation can be determined from Neckel & Klare's (1989) ISM absorption diagrams, or using the observed colour excesses of "normal" stars lying in nearly the same direction as the studied B[e] star. We still note that the observed colour excess obtained from the difference of colour indices $e(b-v)_T = (b-v)_{obs} - (b-v)_*$, actually has three independent components:

$$e(b-v)_T = e(b-v)_{ism} + e(b-v)_{dce} + \Delta(b-v)_g \tag{6}$$

where $\Delta(b-v)_g$ is the "intrinsic" reddening due to continuum absorption and emission phenomena produced by the GCE. It should be remembered that $\Delta(b-v)_g$ does not enter A_v^T in (5), as the physical phenomenon producing it is already taken into account by Δv_g.

2.2.3. *DCE absorption and emission*

UV-visible absorption

Several authors have studied the absorption characteristics of the DCE (cf. Code 1974, Savage et al. 1978, Voshchinnikov et al. 1996, etc.). The effective extinction in the UV-visible spectral range is a combined effect of dust absorption (characterized by an optical depth τ_v) and scattered radiation ($I_v^{scatt.}$) in the DCE. Thus, at the effective wavelength of the magnitude V we have:

$$A_v^{dce} = -2.5 \log\left[e^{-\tau_V} + \frac{I_v^{scatt.}}{I_v^{star+gce}}\right] \quad (7)$$

$$= R_{dce} e(b-v)_{dce} \quad (8)$$

Thus, the scattered radiation, which is not only a function of the optical properties of dust particles but also of the geometry of the DCE, produces a lower effective extinction. Due to this scattered light, absorption A_v^{dce} is a function of the aspect angle (Voshchinnikov et al. 1996). In the same way, changes of the structure of the DCE may partially contribute to the observed variation of A_v^{dce}. The scattered light is also stronger at shorter wavelengths, so that the effective absorption law of the DCE is flatter as compared to those in the ISM. However, at $\lambda \geq 0.25$ Å the effective absorption due to the DCE can be described as a particular ISM absorption law with a somewhat higher value of R (Moujtahid et al. 1998c). Circumstellar and interstellar absorptions by dust can be considered together using Cardelli's et al. (1989) and O'Donnell's (1994) representation of absorption laws. This representation allows us to determine by least-squares the sum of total absorptions $A_v^{ism} + A_v^{dce}$ and the sum of selective absorptions $E(B\text{-}V)_{ism} + E(B\text{-}V)_{dce}$:

$$A_\lambda = a(\lambda)[A_v^{ism} + A_v^{dce}] + b(\lambda)[e(b-v)_{ism} + e(b-v)_{dce}] \quad (9)$$

where a(λ) and b(λ) are tabular functions, the same for all absorption laws. Relation (9) is used once the emission/absorption effects of the GCE are extracted from the observed energy distribution in the studied spectral interval. This correction can be performed using relations (1) and (2). It is worth noting that the absorption in DCE can be wavelength-independent ("grey") if the DCE is clumpy and the individual clouds are optically thick (Boissé 1990, Mitskevitch 1995). In this case use of Cardelli's et al. (1989) representation leads to very high values of R.

IR emission

From about 1μm towards longer wavelengths, the DCE becomes transparent to the radiation field. The energy absorbed in the UV-visible spectral region thus heats the dust and reappears as infrared excess (Savage et al. 1978). This infrared emission by dust was already suggested by Geisel (1970). If there is no other emission source in the infrared, it can readily be realized that for a spherical DCE the integrated energy distribution (dully corrected from ISM absorption) normalized to the energy flux near 1μm is a strong function of the underlying effective stellar temperature.

The balance between absorbed energy in the UV-visible region and reemitted energy in the infrared can be used as an additional "closing" relation. Strong components of IR excess can also be produced by low-temperature free-free emission (Dyck & Milkey 1972, Milkey & Dyck 1973) and by hot, chromospheric-like free-free emission (Vázquez et al. 1993). However, there can be an unknown "loss" or "gain" of emitted energy for the observing direction as a consequence of anisotropic scattering of photons produced by a non-spherical DCE. The relation between the observed amount of UV-visible absorbed energy and that emitted in the IR can be sketched parametrically as follows:

$$\int_0^{1\mu} (absorbed\ energy) d\lambda = P \times \int_{1\mu}^{\infty} (emitted\ energy) d\lambda \qquad (10)$$

where the factor P takes into account the lack of balance between the UV-visible absorbed energy and the observed energy emitted in the IR. Concerning the value of P, Sitko et al. (1981) showed that for 9 of 12 stars with the B[e] phenomenon they studied, the value is $\overline{P} = 1.2 \pm 0.3$. Also, Wang & Hu (1994) using slightly different normalization wavelengths, showed that there is a maximum of HAeBe stars with $P \lesssim 1$.

2.2.4. *Outline of distance determination*

Through the relations written here above we can identify 6 unknown parameters, while the whole set of relations from (1) to (10) actually represent a system of only 4 independent equations. The 6 unknowns are: T_{eff}, luminosity class, D (distance), $E(B\text{-}V)_{dce}$, R_{dce} and P. Two of these parameters then need to be determined by other methods. For example, P can be parametrized so that a reference distance is obtained for $P = 1$ and a range of possible values can be determined for given P_{max} and P_{min}. In spite of this parametrization, consistent solutions are obtained for only a rather narrow interval of P values. Another important parameter that can be determined independently is the effective temperature. This can be done if there is good enough spectrophotometry of the Balmer discontinuity (Zorec et al. 1998). T_{eff} can also be derived by studying the line emission-spectrum excitation (Chkhikvadze 1970, Miroshnichenko 1996). Sometimes, when D > 0.5 pc, $E(B\text{-}V)_{ism}$ becomes constant with distance, so that it may be considered as another known parameter. We note that in this outline the relation $E(B\text{-}V)_{ism} = f(D|l,b)$ enters only as a correction term through Pogson's formula. Once two of 6 parameters are obtained by independent arguments or methods, the remaining 4 are derived by iterating the whole set of equations from (1) to (10). Relation (10) was already used with P = 1 either to determine extinction parameters of B[e] stars (Savage et al.

1978, Sitko et al. 1981) or to discuss the value of T_{eff} (de Freitas Pacheco et al. 1982).

2.2.5. *Results*

Integration over the whole spectral range of the energy distribution requires all wavelengths to be observed simultaneously. However, this can rarely be achieved for B[e] stars, since we deal with data gathered from the literature. From visible photometry, we know that in most cases the flux variations in this spectral region are not very strong. Also, the IR flux variations can be assimilated to the global uncertainties affecting the method. In the far-UV the energy distribution is frequently so strongly absorbed that possible variations (Savage et al. 1978, de Freitas Pacheco et al. 1982) can hardly introduce noticeable differences in the estimation of flux integrated over the whole spectral range. Although such a puzzling recomposition of energy distributions is not the most easiest way of deriving stellar parameters, we obtained the results quoted in Table 3.

TABLE 3. Fundamental parameters, extinctions and distances of some stars with B[e] phenomenon

Star	T_{eff} K	M_{bol} mag	$E(B-V)_{ism}$ mag	A_v^{dce} mag	D pc	Notes
HD 316285	20500	−8.8	1.25	0.57	1770	
HD 163296	11000	+1.16	0.00	0.22	160	
					122^{+17}_{-13}	H
HD 190073	11000	+0.46	0.00	0.43	241	
MWC 314	27500	−10.5	1.55	0.00	1450	
MWC 623	11200	−5.4	0.70	1.58	2400	
MWC 342	11200	−3.7	0.50	2.05	1000	
MWC 349	29000	−8.3	2.30	1.35	970	
MWC 84	34000	−6.9	0.80	2.40	1750	
HD 31648	9000	+1.69	0.05	0.24	150	
					131^{+24}_{-17}	H
HD 51585	28500	−4.0	0.48	0.28	1500	
HD 50138	13400	−2.35	0.09	0.31	300	
					289^{+87}_{-54}	H
HD 45677	22700	−3.17	0.09	1.08	420	
					355^{+226}_{-100}	H
HD 87643	15000	−5.7	0.32	1.46	1450	

Notes: "H" distances measured by HIPPARCOS

2.2.6. *Short comments*

Stars with the B[e] phenomenon presented in Table 3 appear to be of different classes, mainly regarding their evolutionary stage. There are very young objects, others seem to be supergiants, some have characteristics of proto-planetary nebulae or of symbiotic class, while stars considered in the literature as B[e] prototypes cannot be clearly classified (Lamers 1998, Zickgraf 1998). We shall consider them by groups according to their apparent class membership, to comment on how they were studied in the schema of distance determination outlined in the preceding sections.

Unclassified B[e] stars. It is worth mentioning first the group of unclassified objects, for there we find stars which, paradoxically, have been up to now the most studied B[e] stars. In this group we have HD 45677, HD 50138, HD 87643 and MWC 349. HD 45677 and HD 50138 have spectral and polarimetric characteristics of young stellar objects (Schulte-Ladbeck et al. 1992, Bopp 1993, Grady et al. 1993, 1994, Morrison & Beaver 1995, Pogodin 1997, Vaidya et al. 1994, 1995). They are however not associated with molecular clouds as required for HAeBe objects and their evolutionary stage is difficult to determine (Israelian et al. 1996, de Winter & van den Ancker 1997, Jaschek & Andrillat 1997). Conversely, HD 87649 and MWC 349, which are associated with H II regions, are not considered as young objects. Moreover, the latter has a bipolar flow detected in radio frequencies (Cohen et al. 1985, White & Becker 1985). However, its circumstellar disk does not have the structure of an accretion disk (White & Becker 1985). HD 45677 and HD 50138 have spectrophotometric observations in the BCD system (Chalonge & Divan 1952, Burnichon et al. 1971, Zorec et al. 1998). Parameters of the Balmer discontinuity produced by this system are affected neither by the ISM nor by the GCE and DCE absorptions and/or emissions, so that unambiguous photospheric-like visual absolute magnitude, effective temperature and luminosity class of stars underlying the circumstellar envelopes can be determined. The same BCD observations allow us also to determine the second Balmer discontinuity $\Delta\mathcal{D}$. For HD 45677 we adopted v = 7.2 mag as the stellar brightness at the time of the BCD observations (1964) used in the present calculation. Iterations outlined in Sect. 2.2.4 then allow easy estimation of the extinctions affecting the energy distributions of these objects, as well as parameter P in relation (10): P = 1.6 (HD 45677), P = 2.3 (HD 50138). Distances obtained in this paper compare well with those measured by HIPPARCOS. There is abundant spectroscopic literature on HD 87643 (cf. Swings 1974, Carlson & Henize 1979, Surdej et al. 1981, de Freitas Pacheco et al. 1982, 1985, McGregor et al. 1988, Shore et al. 1990, Lópes et al. 1992). Depending on the author, this star was considered to be a P Cyg type star and a nova-like, hot B0V but also a B6-8. For the effective temperature adopted in the present calculation, the star is a bright giant B4II. The star MWC 349 was found to be a binary, where one of the components is a mild Be star with spectral type B0III (Cohen et al. 1985). In this paper we adopted the distance of the Be companion and determined the remaining parameters listed in Table 3.

HAeBe B[e] stars. In this paper we considered three HAeBe stars: HD 31648, HD 163296 and HD 190073 (Thé et al. 1994). They were studied because we expected to have HIPPARCOS parallaxes for all of them, to have an additional test of the outline of distance determination. The effective temperature and an approximate luminosity class (we adopted luminosity class V) of these objects was guessed from the energy distributions obtained by Sitko (1981). This choice leads to distances which agree rather well with those measured by HIPPARCOS.

Proto-planetary nebulae B[e] stars. In this group we consider MWC 84 and HD 51585. The spectral type of MWC 84 was determined by Chkhikvadze (1970) by studying the excitation of the Balmer line emission spectrum. From the spectrophotometry of the star, also obtained by Chkhikvadze, we get $\Delta\mathcal{D} = -0.23$ dex. The spectral type of HD 51585 has been discussed by a number of authors (Andrillat & Houziaux 1973, Klutz & Swings 1977, Arkhipova 1989, Arkhipova & Ikonnikova 1992, Jaschek et al. 1996). The best solution for both stars was obtained using main sequence underlying energy distribution and $P \sim 1$.

VV Cephei-like B[e] stars. This class concerns MWC 342 and MWC 623. These stars have been recognized by Arkhipova & Ipatov (1982) as having an M-type giant companion, although the TiO absorption bands are not clearly identified. Distances of these stars were derived using the spectral types given by Arkhipova & Ipatov (1982).

Supergiant B[e] stars. Depending on the authors, these stars are considered as extreme emission line stars, unclassified B[e] stars or as supergiants of the type of dusty B[e] supergiants of Magellanic Clouds described by Zickgraf (1993), Zickgraf et al. (1996). In this group we consider MWC 314 and HD 316285. MWC 314 was recently studied by Miroshnichenko (1996), who determined the effective temperature we used in the present calculation. In this star there is no evidence for a DCE. The reddening vector of this object scales with $R_{ism} = 2$. The distance quoted in Table 3 is a mean of values obtained using $R_{ism} = 2$ and 3.1. A number of authors have studied HD 316285 and tried to determine its spectral type (Shore et al. 1990, McGregor et al. 1988, van der Veen et al. 1994). The best choice seems to be $T_{eff} = 20500$ K and to consider the star as a supergiant Ia.

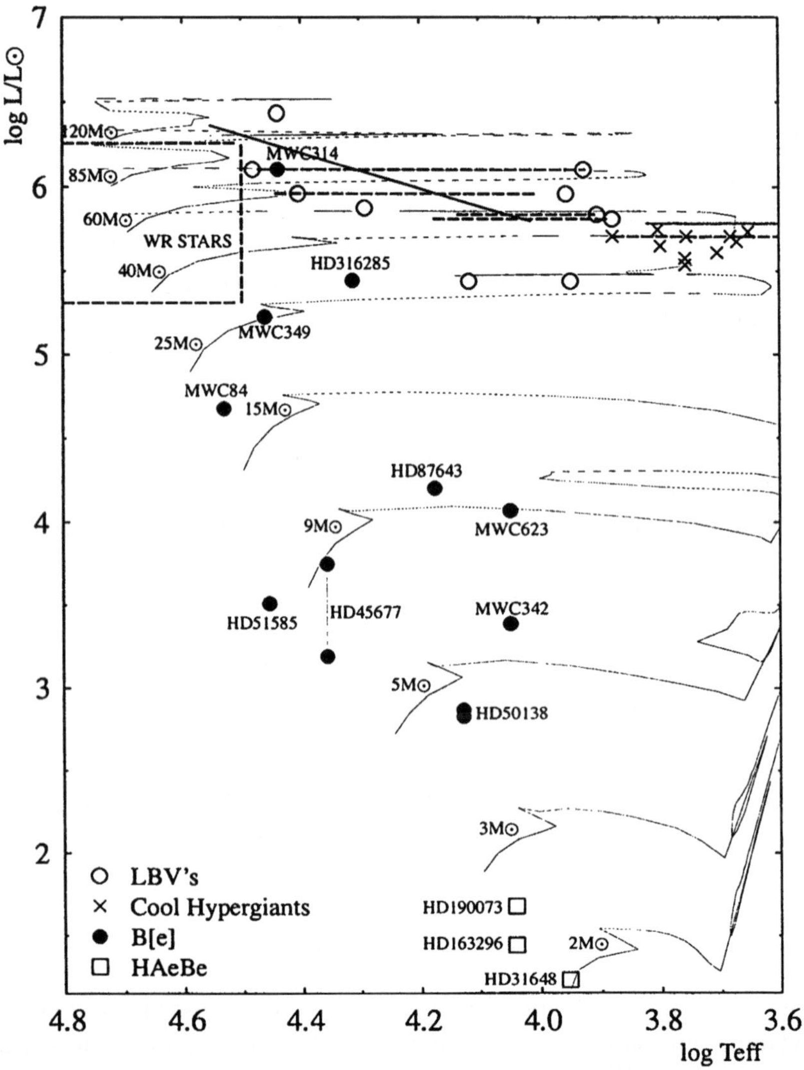

Figure 1. HR diagram of galactic B[e] stars. " • ": B[e] stars; " ○ ": LBV stars; " × ": cool hypergiants; solid lines: temperature-dependent luminosity limits. The location of WR stars is shown.

3. HR diagram of galactic stars with the B[e] phenomenon

Using the parameters listed in Table 3, we obtained the HR diagram shown in Fig. 1, where the evolutionary tracks are from Maeder & Meynet (1988).

For comparison, the location of WR stars, LBV stars and cool hypergiants are also shown. The solid lines represent the temperature-dependent luminosity boundary for hot stars and the luminosity limit for the cool hypergiants (Humphreys 1989). From this diagram we can conclude that stars with the B[e] phenomenon do not have a specific location in the HR diagram: some stars are near the luminosity limit (MWC 314, HD 316285, MWC 349); a number lie among the supergiants (MWC 342, MWC 623); others appear to be on or near the main sequence (HD 50138, MWC 84); while HD 45677 is near the ZAMS and HD 51585 looks like a subluminous star. The position of three HAeBe stars used to check the method of distance determination is also shown. We note, however, that the location of B[e] stars in the HR diagram is apparent and it does not necessarily reveal their actual evolutionary phase. The bolometric luminosity of these stars was calculated from the integrated observed flux and using distances in Table 3. The DCE of B[e] stars may be not spherical, as it is obtained from polarimetric data, and their optical thickness can be rather high. The obtained bolometric luminosity is then proportional to the projected emitting surface of the circumstellar envelope and so it is a function of the aspect angle i: as compared to the actual stellar bolometric luminosity ($L_* = \int L(i)di$), the obtained luminosities are overestimated for stars seen pole-on, while they are underestimated for stars observed equator-on. This phenomenon is shown with HD 45677 and HD 50138 for which we obtained two values of bolometric luminosity: (a) derived from calibrations of photospheric parameters ($\lambda_1, \mathcal{D}$) (upper point in Fig. 1 of HD 45677 and lower point for HD 50138); (b) calculated from integrated observed fluxes.

4. Kinematics of stars with the B[e] phenomenon

Taking advantage of distances discussed above, we can also estimate the galactic velocity of the studied B[e] stars relative to the local standard of rest (peculiar velocities). Peculiar velocities of B[e] stars are affected by a great deal of uncertainties, so that they may be almost meaningless if they are considered individually. We may hope, however, to obtain some information considering them statistically. Using a Monte Carlo technique to simulate the errors affecting simultaneously distances, proper motions and radial velocities, we obtained probability distributions for dispersions of peculiar velocities in each galactic coordinate (Σu: dispersion in the galactocentric direction; Σv: dispersion in the solar motion direction; Σw: dispersion perpendicular to the galactic plane). For comparison we also studied the dispersions of peculiar velocities of "normal" B stars, "classic" Be stars and B supergiants. The most probable dispersions so obtained are given in Table 4. Results of Table 4 show a systematic tendency of velocity dispersions to be higher as the number of stars entering the estimation is lower. This can imply increasing sensitivity of dispersions to errors affecting the observational parameters as the number of stars is smaller, the best example being the dispersions obtained for B[e] stars. Hence, as the velocity dispersions of B[e] stars are probably overestimated, the results obtained suggest that they may belong to Population I. For all types of studied objects we also obtained the most probable mean absolute distance from the galactic plane $\overline{|z|}$. We can also note that there is a small value of $\overline{|z|}$ for B[e] stars. This is possibly due to the fact that genuine HAeBe stars were also taken into account in the present calculation.

TABLE 4. Dispersions of peculiar velocities (in km s^{-1}) of several types of B stars and their mean absolute distances $\overline{|z|}$ (in pc) from the galactic plane.

Star	Sp.T.	Σu	Σv	Σw	$\overline{\|z\|}$	N
	O-B3	9.2	8.1	7.4	72	469
B	B4-A3	7.4	8.3	4.9	45	1589
	O-A2	7.9	8.2	5.6	51	2058
	O-B3	9.3	9.7	9.0	76	124
Be	B4-A3	7.8	7.0	4.4	40	110
	O-A2	8.5	8.6	6.8	58	234
	O-B3	20.6	16.5	21.0	95	65
BI	B4-A3	17.7	16.4	15.6	74	64
	O-A2	18.4	16.2	18.3	84	125
B[e]	all	32.2	11.4	23.8	48	10

5. Galactic distribution of stars with the B[e] phenomenon

Allen & Glass (1975) analysed a sample of 700 stars by near infrared photometry. This sample consists of unresolved planetary nebulae, all types of peculiar Be stars, WR and Of stars, symbiotic and VV Cephei stars. The distribution of the whole sample in galactic longitude shows a strong peak at $\pm 10^o$ around the direction towards the galactic center. This maximum, typical of Population II stars, is due to planetary nebulae. Secondary maxima appear for the Carina and Cygnus direction, which corresponds to stars lying in spiral arms. From this sample Allen & Glass (1975) also isolated the stars with circumstellar dust warmer than about 700 K and obtained for them a distribution in galactic longitude which no longer shares the

Population II stellar distribution. Although this sample still contained a large variety of stellar types, they showed that stars with the B[e] phenomenon considered in isolation share the Population I distribution. Dong & Hu (1991) listed all types of objects with emission lines with colours $(v - [25]) \geq 8$ mag, which also contains almost all known B[e] stars. They showed that their distribution in galactic latitude is strongly concentrated towards the galactic plane and also that: (a) distribution in galactic longitude resembles that of WR stars, (b) it is higher in the direction of the Sagittarius-Carina arm. This suggests that the selected stars belong to Population I.

It can be shown that stars of Table 3 whose distances are $D \leq 1$ kpc share the same location as OB and R associations as well as the H II regions of the Gould belt. This suggests that their age would probably not exceed some 10^7 years (Tejero 1992). Using the diagrams of molecular cloud distribution in the Galaxy (Maddalena et al. 1986, Myers et al. 1986, Dame et al. 1987, Grabelsky et al. 1988, May et al. 1988), it can be shown that stars of Table 3, excepting the HAeBe's, are not associated with molecular clouds. Our program B[e] stars globally follow the space distribution of young star associations and young clusters (Humphreys 1979). They also closely follow the distribution of O9-B5 supergiants (Humphreys 1979) and that of long-period Cepheids (Humphreys 1970).

The rough, first-order kinematic indications obtained in the preceding section and comparison of distributions made in the paragraph above lead us to conclude that stars with the B[e] phenomenon have a galactic distribution resembling that of extreme Population I objects.

6. Conclusions

We have presented a short review of indirect methods of distance estimation for stars with the B[e] phenomenon. Although these methods are currently used, they are highly uncertain. The B[e] phenomenon is seen in objects which are apparently in different evolutionary phases or are of different nature. As the nature of stars underlying the circumstellar envelopes that produce the characteristic features of the B[e] phenomenon is almost unknown, reliable methods of distance determination, such as the spectroscopic method, are difficult to generalise. However, our present knowledge of the long-term photometric behaviour of B[e] stars, their characteristic energy distribution from the UV to the far IR and the distinction made between different emission and absorption components affecting the visual magnitude V, prompted an outline of distance determination for these objects. This schema seems to be consistent, as the results obtained for some HIPPARCOS B[e] targets are concordant. Sometimes, results may depend rather strongly on the chosen stellar effective temperature and/or luminosity class. To overcome this difficulty, it would be useful to study in some detail the energy distribution of these objects around the Balmer discontinuity, as it leads to a less ambiguous choice of fundamental stellar parameters. Quantitative studies of excitation of the emission spectrum would also help to determine the effective temperature of stars underlying the complex envelopes of B[e] stars. Simultaneous energy distributions would enable better study of the balance of absorbed and emitted energy by the circumstellar envelope. This balance would be important to determine the actual efficiency of different mechanisms able to produce the IR excess.

The distances obtained were also used to infer the bolometric luminosity of the studied stars, their peculiar galactic velocitites and to study their galactic distribution. The bolometric luminosity of B[e] stars, obtained by integrating the observed fluxes, is aspect angle-dependent and leads to locations of stars in the HR diagram that do not necessarily correspond to actual evolutionary stages. The peculiar velocities of B[e] stars can be considered only statistically, because they are affected by strong observational uncertainties. Nevertheless, the values obtained for their dispersions seem to be low, so that on kinematical grounds, B[e] stars may be Population I objects. Moreover, their galactic distribution resembles that of extreme Population I stars.

Acknowledgments

The names of groups of B[e] stars used in subsection 2.2.6 were inspired during discussion hold with Drs H. Lamers, F.J. Zickgraf, D. de Winter and L. Houziaux.

References

Allen D.A., Glass I.S. 1975, *MNRAS*, **170**, 579
Allen D.A., Swings J.P. 1976, *A&A*, **47**, 293
Andrillat Y., Houziaux L. 1973, *Mem. Soc. Roy. Sc. Liège*, **6th ser.**, vol. V, p.377
Arkhipova V.P. 1989, *SvAL*, **15**, 400
Arkhipova V.P., Ikonnikova N.P. 1992, *SvAL*, **18**, 418
Arkhipova V.P., Ipatov A.P. 1982, *SvAL*, **8**, 288
Ballereau D., Chauville J., Zorec J. 1995, *A&AS*, **111**, 423
Bergner Yu.K., Miroshnichenko A.S., Yudin R.V. et al. 1995, *A&AS*, **112**, 221
Blitz L., Fich M., Stark A.A. 1982, *ApJS*, **49**, 183
Boissé P. 1990, *A& A*, **228**, 483
Bopp B.W. 1993, *IBVS* 3834
Burnichon M.L, Chalonge D., Divan L. et al. 1971, *J. Obs.*, **50**, 391
Cardelli J.A., Clayton G.C., Mathis J.S. 1989, *ApJ*, **345**, 245
Carlson E.D., Henize K.G. 1979, *Vistas in Astronomy*, **23**, 213
Chalonge D., Divan L. 1952, *Ann. Astrophys.*, **15**, 201
Chkhikvadze Y.N. 1970, *Astrofizika*, **6**, 65

Code A.D. 1974, In: Interstellar Dust and Related Topics, (eds.) Geenberg J.M., van de Hulst H.C., IAU Symp. N52, Reidel Dordrecht, p. 505
Cohen M., Bieging J.H., Dreher J.W. et al. 1985, *ApJ*, **292**, 249
Dame T.M., Ungerechts H., Cohen R.S. et al. 1987, *ApJ*, **322**, 706
Dong Y., Hu J. 1991, *Chin. Astron. Astrophys.*, **15**, 275
Dyck H.M., Milkey R.W. 1972, *PASP*, **84**, 501
de Freitas Pacheco J.A., Faria Lópes D., Landaberry S.C. et al. 1985, *A&A*, **152**, 101
de Freitas Pacheco J.A., Girla D.P., Pottasch S.R. 1982, *A&A*, **108**, 117
Geisel 1970, *ApJ*, **161**, L105
Grabelsky D.A., Cohen R.S., Bronfman L. et al. 1988, *ApJ*, **331**, 181
Grady C.A., Bjorkman K.S., Shepherd D. et al. 1993, *ApJ*, **415**, L39
Grady C.A., Pérez M.R., Thé P.S. 1994, ASP Conf. Ser. **62**, 409
Herbig G.H. 1975, *ApJ*, **196**, 129
Herbig G.H., Leka K.D. 1991, *ApJ*, **382**, 193
Herbst W. 1975a, *AJ*, **80**, 212
Herbst W. 1975b, *AJ*, **80**, 503
Hobbs L.M. 1974, *ApJ*, **191**, 381
Humphreys R. 1970, *AJ*, **75**, 602
Humphreys R. 1979, In: *The Large-Scale Characteristics of the Galaxy*, IAU Symp. No **84**, (ed.) Burton W.B., p. 93
Humphreys R. 1989, In: *Physics of luminous blue variables*, (eds). Davidson K., Moffat A.F.J., Lamers H.J.G.L.M., Kluwer Acad. Publ., p.3
Israelian G., Friedjung M., Graham J. et al. 1996, *A&A*, **311**, 643
Jaschek C. 1998, In: *Workshop on B[e] stars*, (eds.) Hubert A.M., Jaschek C., this issue
Jaschek C., Andrillat Y. 1997, *A&AS*, preprint
Jaschek C., Andrillat Y., Jaschek M. 1996, *A&AS*, **117**, 281
Jaschek M., Egret D. 1982, CDS-SP No. **4**, Observatoire de Strasbourg
Jaschek M., Slettebak A., Jaschek C. 1981, *Be stars Newsletter* No. **4**, p. 9
Klutz M., Swings J.P. 1977, *A&A*, **56**, 143
Lamers H.J.G.L.M. 1998, In: *Workshop on B[e] stars*, (eds.) Hubert A.M., Jaschek C., this issue
Lópes D.F., Damineli Neto A., de Freitas Pacheco J.A. 1992, *A&A*, **312**, 941
Maddalena R.J., Morris M., Moscowitz J. et al. 1986, *ApJ*, **303**, 375
Maeder A., Meynet G. 1988, *A&AS*, **76**, 411
Mathis J.S. 1990, *Ann. Rev. Astron. Astrophys*, **28**, 37
May J., Murphy D.C., Thaddeus P. 1988, *A&AS*, **73**, 51
McGregor P.J., Hyland A.R., Hillier D.J. 1988, *ApJ*, **324**, 1071
Mihalas, D., Binney J. 1982, *Galactic Astronomy*, Freeman & Co.
Milkey R.W., Dyck H.M. 1972, *ApJ*, **181**, 833
Miroshnichenko A.S. 1996, *A&A*, **312**, 941
Mitskevitch 1995, *A&A*, **298**, 219
Morrison N.D., Beaver M. 1995, *BAAS*, **27**, 825
Moujtahid A., Zorec J., Hubert A.M. et al. 1997, *A&AS*, in press
Moujtahid A., Zorec J., Hubert A.M. 1998a, In: *Workshop on B[e] stars*, (eds.) Hubert A.M., Jaschek C., this issue
Moujtahid A., Zorec J., Hubert A.M. 1998b, In: *Workshop on B[e] stars*, (eds.) Hubert A.M., Jaschek C., this issue
Moujtahid A., Zorec J., Hubert A.M. et al. 1998c, *A&A*, in preparation
Myers P.C., Dame T.M., Thaddeus P. et al. 1986, *ApJ*, **301**, 398
Neckel Th., Klare G. 1989, *A&AS*, **42**, 251
O'Donnell J.E. 1994, *ApJ*, **422**, 158
Oudmaijer R. 1998, In: *Workshop on B[e] stars*, (eds.) Hubert A.M., Jaschek C., this issue
Pogodin M.A., 1997, *A&A*, **317**, 185
Reddish V.C. 1967, *MNRAS*, **135**, 251

Savage B.D., Wesselius P.R., Swings J.P. et al. 1978, *ApJ*, **224**, 149
Scheffler H. 1982, *Landolt-Börnstein*, GVI, Vol. **2**c
Schulte-Ladbeck R.E., Shepherd D.S., Nordsieck K.H. et al. 1992, *ApJ*, **401**, L105
Sitko M.L. 1981, *ApJ*, **247**, 1024
Sitko M.L., Savage B.D. 1980, *ApJ*, **237**, 82
Sitko M.L, Savage B.D., Meade M. 1981, *ApJ*, **246**, 161
Sharpless S. 1959, *ApJS*, **4**, 257
Shore S.N., Brown D.N., Bopp B.W. et al. 1990, *ApJS*, **73**, 461
Sneden C., Gehrz R.D., Hackwell J.A. et al. 1978, *ApJ*, **223**, 168
Snow Th.P., York D.G., Welty D.E. 1977, *AJ*, **82**, 113
Surdej A., Surdej J., Swings J.P. et al. 1981, *A&A*, **93**, 285
Swings J.P. 1974, *A&A*, **34**, 333
Tejero F.C. 1992, *PhD Thésis*, Univerity of Barcelona, Spain
Thé P.S., de Winter D., Pérez M.R. 1994, *A&AS*, **104**, 315
Vaidya A., Schulte-Ladbeck R.E. 1995, *BAAS*, **27**, 825
Vaidya A., Schulte-Ladbeck R.E., Bjorkman K.S. 1994, *BAAS*, **26**, 933
Vázquez A.C., Cidale L.S., Ringuelet A.E. 1993, *ApJ*, **419**, 286
van der Veen W.E.C.J., Waters L.B.F.M., Trams N.R. et al. 1994, *A&A*, **285**, 551
Vega E.I., Muzzio J.C., Feinstein A. 1986, *Rev. Mexicana Astron. Astrof.*, **13**, 33
Voshchinnikov N.V., Molster F.J., Thé P.S. 1996, *A&A*, **312**, 243
Wang J., Hu J. 1994, *Chin. Astron. Astrophys.*, **16**, 293
Welty D.E., Hobbs L.M., Kulkarni V.P. 1994, *ApJ*, **436**, 152
Westerlund B.E. 1987, *A&AS*, **70**, 311
White R.L., Becker R.H. 195, *ApJ*, **297**, 677
de Winter D., van den Ancker M.E. 1997, *A&AS*, **121**, 275
de Winter D., van den Ancker M.E., Pérez M.R. et al. 1996, *A&AS*, **119**, 1
Zickgraf F.-J. 1993, In: *Lecture Notes in Physics*, **416**, (eds.) Baschek B., Klare G., Lequeux J., p.257
Zickgraf F.-J., Humphreys R.M., Lamers H.J.G.L.M. et al. 1996, *A&A*, **315**, 510
Zorec J., Briot D. 1991, *A&A*, **245**, 150
Zorec J., Briot D. 1997, *A&A*, **318**, 443
Zorec J., Moujtahid A., Ballereau D., Chauville J. 1998, In: *Workshop on B[e] stars*, (eds.) Hubert A.M., Jaschek C., this issue

Discussion

Houziaux: For HD 51585, with T_{eff} = 28 500 K and M_{bol} = -4.0 it would seems that $R/R_{\odot}$ would be only ≈ 2.0 which means that this star must be far from the main sequence. Is it true?

Zorec: The M_{bol} of B[e] stars were obtained from integrated observed fluxes. Assuming that the circumstellar envelope is ellipsoidal, the anisotropic scattering of photons will produce an apparent bolometric luminosity wich is aspect angle-dependent. Assuming the envelope has uniform brightness and an ellipticity E < 1, the luminosity loss for equator-on seen stars is $\Delta L/L \propto E^{-2/3}$ for stars seen pole-on. If there is now knowledge of E and the aspect-angle, the apparent bolometric luminosities of B[e] stars lead to meaningless stellar radii.

Lamers: One of the effects that you take into account in modelling the energy distribution is scattering by circumstellar gas. But this effect must depend on the geometrical extent of the CS gas cloud compared to the aperture of the photometer or spectrograph. Did you assume that the scatteried radiation is all within the aperture of the photometer?

Zorec: Yes, I supposed so.

Zickgraf: Did you try to calculate kinematical distances from radial velocities?

How do they compare to the distance they obtain from the continuum method?

Zorec: Only mean values of kinematical distances have some meaning, as one can assume that the mean of peculiar velocities is zero. Otherwise, the individual kinematical distances are affected by a spurious component $N_{Pec}/A\sin 2l \cos^2 b$. In our set of B[e] stars, $<N_{Pec}> \neq 0$, so that $<D>_{Dyn} \neq <D>_{Method}$. For HD 45677 $N_{Pec} \sim 0$ and $D_{Dyn} \sim 0.39$ Kpc, while $D_{Hipparcos} = 0.36$ Kpc.

ABSOLUTE MAGNITUDES FOR SOME B[e] STARS FROM HIPPARCOS

D. BRIOT AND N. ROBICHON

DASGAL, Observatoire de Paris
5, place Jules Janssen, 92195 Meudon CEDEX

1. Introduction: the situation before Hipparcos

Before the measurements of the Hipparcos satellite, only the parallaxes of very few B stars were measured: B stars, which are intrinsically bright, are scarce in the solar neighbourhood and the great majority of them are out of range of ground-based parallax measurements. The parallaxes of Be stars were only determined by indirect methods. However, because their absolute magnitude is a clue for the understanding of the Be phenomenon which is up to now very badly explained, numerous studies were devoted to the determination of this parameter, since Merrill (1933). Extensive references can be found in Zorec and Briot (1991). At it will be seen below, because of the measurements of the parallaxes for large samples of B and Be stars, the results of Hipparcos bring new insight on the absolute magnitudes of Be stars as compared with the absolute magnitude of B stars. A preliminary study was already published in Briot et al. (1997).
As far as the absolute magnitudes of galactic B[e] stars are concerned, everything is to be discovered both in observation and in theory. It is only the study of the position of these stars in the HR diagram which will enable us to clarify their evolutionary status. The aim of the present study is to determine the absolute magnitudes of B[e] stars comparison with both the absolute magnitudes of B and Be stars, all from Hipparcos measurements. It should be specified that the absolute magnitude of Be and B[e] stars studied here, is the "total" absolute magnitude, that is to say, the absolute magnitude of the central star, plus the circumstellar regions.

A. M. Hubert and C. Jaschek (eds.), B[e] Stars, 47–52.

2. Analysis and results

Our observational program, as was proposed in 1982 for the Hipparcos input catalogue, contained around 14000 B stars, with and without emission. Actually, about 5300 of these stars were observed by Hipparcos. We eliminated from the sample studied the stars for which the parallax obtained is negative, as well as double stars. The giants were excluded too, and this study is devoted to the stars of luminosity class V or IV. Be stars are found from the Jaschek and Egret catalogue (1982), as well as from the B stars noted with emission in the Hipparcos catalogue. So the sample studied contains 2077 B stars with no detected emission, 213 Be stars and 4 B[e] stars.

HD	HIP	spectral type	π (mas)	σ_π/π	distance (pc)	M_V (mag)
45677	30800	B2 V	2.82	0.39	355	-0.02
50138	32923	B5 V	3.46	0.23	289	-0.88
161114	86777	B0 V	1.83	0.60	546	-1.97
163296	87819	B9 V	8.19	0.12	122	1.13

TABLE 1. *Distances and absolute magnitudes of B[e] stars*

We established a HR diagram for B and Be stars, in which we plotted the individual B[e] stars. We determined for this purpose the mean values of absolute magnitudes as a function of the spectral type for the B and Be stars.
The interstellar absorption Av is computed using the tridimensional model of Arenou et al. (1992).
The results are shown on figure 1. The error bars are determined taking into account errors in the parallaxes as well as errors in Av.
We obtained :

- for the Be stars : the general trend is an over-luminosity of Be stars as compared with B stars of the same spectral type, this over-luminosity being larger for the later spectral types.
- for the B[e] stars : the 4 B[e] stars for which Hipparcos astrometric data exist are located under the mean sequence of B stars. So they are under-luminous.

The numerical results obtained from Hipparcos observations for these 4 stars are presented in table 1, i.e. parallaxes, errors on the parallaxes, distances and absolute magnitudes.

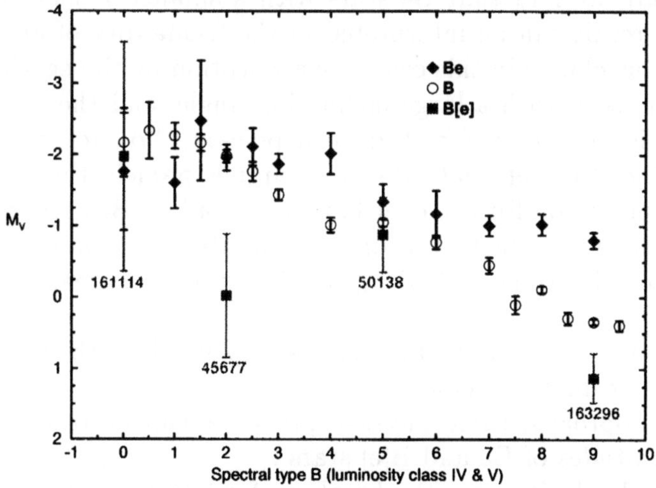

Figure 1. Observational HR diagram for B, Be and B[e] stars.

3. Discussion and physical interpretations

The spectral types used for B and Be stars are those of the Hipparcos Input Catalogue. Actually, as the results presented here are mean results per spectral type, and as no bias has been detected in the spectral type determinations, the mean values should not be deeply modified.
The spectral type of B[e] stars is particularly difficult to determine. For example, determinations of the spectral type of the star HD 50138 range from B2 to B9, according to the lines considered (Jaschek et al. 1991).
All positive parallaxes are considered here whatever the uncertainty in the measurements. Certainly the results obtained for the hottest Be stars (B0-B1) are less reliable than for the other spectral types, firstly because of the smallness of these samples, secondly because the main sequence is significantly broader than for later spectral types, and lastly because stars of these spectral types are very distant and so the accuracy of their parallaxes determination is rather poor. It is important to note that the results obtained for the other spectral types remain the same even when only stars with $\sigma_\pi/\pi < 20\%$ are considered. Moreover a preliminary study taking into account the Malmquist bias gives similar results.

In a preliminary study, we considered the absolute magnitudes of B2e stars with $\sigma_\pi/\pi < 30\%$ as a function of $V \sin i$ (Briot et al. 1997). Studying

separately the strong emission stars with a larger rotational velocity and the weak emission stars with a smaller rotational velocity, as it was shown in Briot (1986), we find that Be stars with a smaller $V \sin i$ are more luminous. This result can be interpreted in the framework of an anisotropic circumstellar envelope : in this case, the absorption by the envelope is maximum for stars seen with a large inclination angle, and the re-emission by the envelope is maximum for stars seen pole-on. No such result was obtained from the study of a late Be stars sample, namely the B9e stars.
A possible explanation for the over-luminosity of late Be stars can be some effect due to their very fast rotation. Indeed late Be stars rotate very near their break-up velocity ($V \simeq 0.9V_{crit}$ for B9e stars), whereas early Be stars do not do so.
The under-luminosity of B[e] stars can be explained by the strong absorption due to their dusty envelopes.
So, in a first approach, three physical processes can be responsible of the absolute magnitudes of Be and B[e] stars :

- Rotational velocity near the break-up for late Be stars;
- Inclination angle of an anisotropic envelope for early Be stars;
- Absorbing envelopes for B[e] stars.

4. Conclusion

As said above, a more detailed study of absolute magnitudes of B, Be and B[e] stars is in progress, in which we shall take into account :

- A more accurate determination of spectral types, particularly for Be and B[e] stars;
- The Malmquist bias;
- The Lutz and Kelker bias;
- A correction for the undetected Be stars in the sample of B stars.

Absolute magnitudes of Be and B[e] stars so determined will give new constraints to model these stars.

Acknowledgments

Thanks are due to Frédéric Arenou for helpful comments and to Michael Friedjung for improving the presentation of this paper

References

Arenou F., Grenon M. and Gomez, A. (1992), *A&A*, **258, 104**
Briot D. (1986), *A&A*, **163, 67**

Briot D., Robichon N. and Hubert A.M. (1997), *In : Hipparcos Venice'97*, **ESA SP-402, 319**
Jaschek M., Jaschek C., Andrillat Y. and Houziaux I. (1992), *MNRAS*, **254, 413**
Jaschek M., Egret D. (1982), *Catalogue of Stellar Groups*, **SP. Publ. CDS 4, Obs. Strasbourg**
Merrill P.W., (1933), *PASP*, **45, 198**
Zorec J., Briot D. (1991), *A&A*, **245, 150**

Discussion

Owocki: Why are the Be stars systematically brighter than ordinary B stars of the same spectral type? That is, what physical interpretation do you give for this empirical result?
Briot: The most likely explanation for this observational result is probably the faster rotation of Be stars. All the characteristics of emission of Be stars decrease on average with the temperature of the central star. However, the rotational velocity of late Be stars is very near the break-up velocity for these stars, which is not the case for early Be stars.
Miroshnichenko: Did you take into account in your Mv determination that HD 161114 = XX Oph is a binary star with a late-type component?
Briot: Actually, I did not consider that HD 161114 is a binary star, but the error bar is very large for this star. If we take into account a late-type companion, the B[e] star is shifted downwards in the HR diagram, which would be in agreement with our general result that, on average, the B[e] stars are less luminous than the B stars.
Zickgraf: HD 45677 is photometrically variable. J. Zorec has just shown the light-curve. Does the absolute M_V you give correspond to the faint state of the light-curve? This would explain the difference as compared to the M_V given by Zorec which is about 2 magnitudes brighter?
Briot: The V magnitude given by Hipparcos is the mean magnitude. Our results and the one given by Zorec should coincide.
Oudmaijer: When you derived absolute magnitudes of normal B stars, you used stars with $\sigma_\pi/\pi < 20\%$ to avoid the Lutz-Kelker bias. The errors fot the B[e] stars are much larger. Does this have any implication on the resulting absolute magnitudes?
Briot: Actually the results presented here concerned all the B and Be stars with positive parallaxes and not detected as double by Hipparcos. We also verified that these general results remained the same when only stars with $\sigma_\pi/\pi < 20\%$ were considered. I wish also to point out that absolute magnitudes of B and Be stars are mean results per spectral type, whereas absolute magnitude of B[e] stars are results for individual stars. So the error bars have different meanings. Moreover, the mean sequences for B and Be stars are wider than expected, specially for early type stars.

Up to now the result concerning the B[e] stars is only a general trend. We are now applying more detailed statistical treatments taking into account different biases. However, B and Be stars probably share the same biases in the present study and their differential behaviours should not change when these biases are corrected.

Lamers: I am interested in your result that the Be stars are brighter in Mv than the normal B stars of the same spectral type and luminosity class. We found the same effect when we compared the slowly and rapidly rotating early-B stars with Hipparcos. We explained this effect in terms of errors in the luminosity classes assigned to rapidly rotating stars. This is confirmed by detailed studies of a few stars with model atmospheres (see Lamers et al. 1997 A&A Letters, in press).

Cassinelli: George Collins once showed that rapid rotation of a B star makes it appears later in spectral type. So the B9 stars in your plot could perhaps be B6 stars but affected by their rapid rotation. Also one is more likely to see a star nearly equator-on than pole-on so there can be a preferential shift toward later spectral type.

Waters: The fact that late type Be stars are more luminous than similar non-Be stars can be due to the effect of rapid rotation on spectral type, but then these stars should all be edge-on. A pole-on Be star with the same spectral type would have the luminosity of the non-Be stars.

TYCHO PHOTOMETRY OF SOME B[e] STARS

M. FLOQUET
URA 335 du CNRS, DASGAL, Observatoire de Paris-Meudon 92195 Meudon Cedex, France

J.L. HALBWACHS
Observatoire de Strasbourg 11 rue de l'Université, 67000 Strasbourg, France

AND

A.M. HUBERT
URA 335 du CNRS, DASGAL, Observatoire de Paris-Meudon 92195 Meudon Cedex, France

1. Introduction

The Tycho experiment was launched with the Hipparcos satellite and provided two-colour photometry B_T and V_T for more than one million stars brighter than V_T =11.5 mag (ESA 1997, ESA SP-1200). These two filters are roughly centered on the B and V Johnson filters. A simple linear transformation exists between Tycho B_T and V_T magnitudes and B and V magnitudes in the Johnson photometric system (see ESA 1997). The Tycho Epoch Photometry Annex A contains the B_T and V_T magnitudes measured for each Tycho observation of the known variable stars. In the present paper, these measurements are corrected for the bias described in ESA SP–1200, Vol. 4, chapter 9 (pg 133).
Table 1 provided Tycho photometry for 5 stars known to present spectral characteristics of the so-called B[e] stars. In this table can be found for each star of our sample, mean values of B_T and V_T given in the Tycho catalogue (ESA 1997, ESA SP-1200), the mean value of errors of all individual magnitudes over all the data set σB_T and σV_T, the differences between extrema values in each filter, ΔB_T and ΔV_T, the mean value of $(B_T - V_T)$ and $\Delta V_T/\Delta(B_T - V_T)$.

A. M. Hubert and C. Jaschek (eds.), B[e] Stars, 53–54.

TABLE 1. Tycho photometry

HD, name	B_T	σB_T	ΔB_T	V_T	σV_T	ΔV_T	$(B_T - V_T)$	$\Delta V_T/\Delta(B_T -$
45677	8.105	0.052	0.344	8.052	0.063	0.323	0.041	-0.738
FS CMa								
50138	6.612	0.051	0.096	60585	0.057	0.119	0.030	-0.621
94878	9.420	0.154	0.561	8.828	0.149	0.502	0.575	-0.710
GG Car								
161114	10.050	0.274	0.617	9.041	0.120	0.407	0.980	-0.318
XX Oph								
316285	11.147	0.350	1.300	9.362	0.136	0.504	1.553	-0.151

2. Results

- HD45677: This star is generally given as the prototype of B[e] stars. It is known to show large variations in light up to 2.0 mag in the V filter. Tycho photometry displays a variation in B_T and V_T of roughly 0.3 mag. The V_T-curve coincides very well with the V-curve published by de Winter & van den Ancker (1996, A&AS 121, 275) (see also Miroshnishenko, this volume).

- HD50138: This star is also known to present variations in light up to 0.4 mag in the V-filter. Tycho photometry does not display variation larger than 3σ in B_T and V_T. This epoch corresponds to the flat part of the light curve (see also Miroshnishenko, this volume).

- HD94878: This star is an eclipsing binary with a period P=62.09 days. The light curve amplitude of this binary is 0.4 mag in V. B_T and V_T curves obtained with the ephemerid given by Gosset et al. (1984, A&AS 55, 411) are in good agreement with the curves published by these authors, but the amplitude is of the same order as 3σ.

- HD161114: This star has shown in the past irregular fluctuations between 9.1 and 11.1 in the V magnitude. Tycho data correspond to the brighter part of the V-curve but do not show variations larger than 3σ.

- HD316285: The variations observed in B_T and V_T are slightly larger than 3σ. We have observed 2 small bursts in V_T at HJD 2448186 and 2448672 with an amplitude of 0.40 and 0.35 mag respectively. These bursts cannot be explained by interference with other stars in the field.

In conclusion, Tycho photometry has confirmed light variability in a small sample of B[e] stars. The most important findings is the general trend for stars to be redder when they are brighter.

FUNDAMENTAL PARAMETERS OF TWO B[e] STARS: HD 45677 AND HD 50138

J. ZOREC[1], A. MOUJTAHID[2], D. BALLEREAU[2], J. CHAUVILLE[2]
[1]*Institut d'Astrophysique-CNRS, 98bis, bd. Arago, F-75014 Paris*
[2]*Observatoire de Paris-Meudon, F-92195 Meudon Principal. Cedex*

1. Observations and basic parameters

HD 45677 and HD 50138 were observed sporadically in the BCD system (1) from 1959 to 1980. The photospheric component of the Balmer discontinuity (BD) in both stars has remained unchanged for the whole observing period. The uneven energy distribution near the BD of HD 45677 makes difficult its luminosity class determination. Hence, we derived two extreme luminosity classes. The $(\lambda_1, \mathcal{D})$ (mean position and value of the BD respectively) and the corresponding fundamental parameters of the studied B[e] stars are given in Table 1:

TABLE 1. Fundamental parameters of HD 45677 and HD 50138

Star	$\mathcal{D}$ [dex]	λ_1 [Å]	Sp.Ty.	T_{eff} [K]	$M_V(\lambda_1, \mathcal{D})$ [mag]	$M_{bol}(\lambda_1, \mathcal{D})$ [mag]
HD 45677	0.143	3750	B2III-IV	21200	-3.00	-5.1
	0.143	3760	B2V	22700	-2.20	-4.6
HD 50138	0.294	3743	B6III-IV	13400	-1.40	-2.3

Distances were determined using the method outlined in (3). The spectroscopic distances (D_{Sp}) and those measured by HIPPARCOS (D_{hipp}), interstellar (ISM) reddening, absorption by the circumstellar dust (CSD) and bolometric magnitude derived from integrated fluxes are given in Table 2

From Na I D lines (equivalent widths W(D1) = 0.32 [Å], W(D2) = 0.26 [Å]) and the classic statistical relations (2), we derived for HD 45677:

A. M. Hubert and C. Jaschek (eds.), B[e] Stars, 55–56.

TABLE 2. Distances and extinctions of HD 45677 and HD 50138

Star	D_{Sp} [pc]	D_{hipp} [pc]	$E(B\text{-}V)_{ISM}$ [mag]	A_V^{CSD} [mag]	$M_{bol}(\mathcal{F})$ [mag]
HD 45677	606		0.09	1.28	-3.5
	420	355^{+226}_{-100}	0.08	1.25	-2.9
HD 50138	303	289^{+87}_{-54}	0.09	0.59	-2.4

D = 725 [pc] and $E(B\text{-}V)_{ISM}$ = 0.25 mag. The equivalent widths of ISM molecular bands of HD 50138 observed in CASLEO (Argentina) (W(4430) = 0.056 [Å], W(5789) = 0.018 [Å] and W(6284) = 0.410 [Å]) lead to E(B-V) = 0.03, < 0 and 0.29 [mag] respectively. As $E(B\text{-}V)_{total}$ = 0.27 [mag], we see that distances and selective reddenings determined from statistical relations are strongly uncertain.

2. Discussion

By comparing the spectroscopic and trigonometric (HIPPARCOS) distances, we can say that the spectral type of HD 45677 should be: **B2V**. For this star there is a difference $\Delta M_{bol} \simeq 1.7$ mag between both M_{bol} determinations (Table 1 vs. Table 2). Two simple explanations can be given: (*i*) the star has a flattened circumstellar envelope (CE) of dust and gas with ellipticity E $\simeq$ 0.1 that produces a non isotropical scattering of light; (*ii*) the stellar radius is half of that corresponding to a normal B star with $(\lambda_1, \mathcal{D})$ parameters. However, as the absorbed energy in the visible-ultraviolet is about 1.6 times that of the energy emitted in the IR, a disc-like structure of the dusty envelope seen edge-on would be the most plausible explanation. The radius of HD 45677 inferred from the $(\lambda_1, \mathcal{D})$ calibrations is $R/R_\odot = 4.9$, which corresponds to a mean main sequence star with $M/M_\odot \simeq 8.8$.

In HD 50138 we see that $M_{bol}(\lambda_1, \mathcal{D}) \simeq M_{bol}(\mathcal{F})$. This may suggest that the star is in a giant-like stage and that its dusty CE is nearly spherical.

References

Chalonge, D., Divan, L. (1952), *Ann. Ap.*, **15**, p.201
Scheffler, H. (1982), *Landolt-Börnstein*, G.VI, Vol. **2c**, p.55
Zorec, J. (1997), *this issue*

ULTRAVIOLET EXTINCTION OF CIRCUMSTELLAR DUST IN HD 45677

A. MOUJTAHID[1], J. ZOREC[2], A.M. HUBERT[1]
[1] *Observatoire de Paris-Meudon, F-92195 Meudon Principal. Cedex*
[2] *Institut d'Astrophysique*-CNRS, *98bis, bd. Arago, F-75014 Paris*

1. Method

The far-UV extinction towards a B[e] star has at least two components: interstellar and circumstellar. The observed absorption in the visible and UV range due to circumstellar dust is a medley effect of absorption and scattered stellar radiation light (4), so that its distribution is flatter as it would correspond to dust particles described by high values of $R_V = A_V/E(B-V)$ (ratio of total to selective extinction). The aim of this paper is to study whether it is possible to model the observed far-UV (from 0.12μm to 0.31μm) absorption A_λ of B[e] stars with a unique biparametric representation as given by (1): $A_\lambda = a(\lambda) \times A_V + b(\lambda) \times E(B-V)$, where $A_V = A_V^{ISM} + A_V^{CSD}$, $E(B-V) = \mathrm{E(B\text{-}V)}_{ISM} + E(B-V)_{CSD}$ (ISM stands for interstellar component and CSD for circumstellar dust) and a(λ), b(λ) are universal interpolatory functions. We used archive IUE data for HD 45677. As comparison star we used HD 64802 (B2V, E(B-V) = 0.05).

2. Results and discussion

From the interstellar absorption in the direction of HD 45677 and its HIPPARCOS distance, we derived $A_V^{ISM} = 0.25$. Assuming that for the ISM absorption we may use $R_V = 3.1$, we derived the extinction parameters given in Table 1, which lead to the following conclusions:

(*i*) Neither the total A_λ nor the A_λ^{CSD} can be satisfactorily represented using a single value of R_V in the whole $\lambda\lambda 0.12 - 0.31$ μm interval.

(*ii*) The general slopes of A_λ would correspond to an intermediate value of R_V, while the small absorption in the bump at 0.22 μm would likely correspond to a high R_V.

A. M. Hubert and C. Jaschek (eds.), B[e] Stars, 57–59.

The values of A_V and R_V given in Table 1 that represent the bump at 0.22 μm were derived by searching for a fit of the observed curves A_λ in $\lambda\lambda$ 0.22 – 0.31 μm interval only. By neglecting the presence of the bump at 0.22 μm, we obtain the values R_V^{slope} also given in Table 1.

TABLE 1. Absorption parameters for HD 45677

Dates	A_V	R_V	A_V^{CSD}	R_V^{CSD}	R_V^{slope} ISM+CSD	R_V^{slope} CSD
20.09.78	2.39	6.43	2.14	7.35	4.12	4.30
21.02.92	2.03	5.83	1.78	6.64		
27.02.92	1.50	5.08	1.26	5.82	3.78	3.92
22.12.92	2.05	5.99	1.80	6.87	3.88	4.04
20.09.95	1.90	5.61	1.66	6.39	3.90	4.06

The difficulty of obtaining a consistent representation of the observed extinction curves A_λ in HD 45677 over the entire wavelength range from 0.12μm to 0.31μm with a single value of R_V reveals the presence of other components involved in radiation absorption in the dusty envelope as compared to those produced in the interstellar medium. The simplest explanation is the presence of scattered stellar radiation, which produces a lower and flatter mean effective absorption in the whole spectral range (4). The scattered stellar radiation in dust shells is dependent on the particle optical properties and on the geometry of the circumstellar envelope. It thus seems likely to suppose that the observed variations of A_V^{CSD} and R_V in Table 1 reflect not only possible changes of optical properties of dust particles but also variations of the geometrical structure of the circumstellar envelope. It is worth noting that in the particular case of clumpy circumstellar envelopes, the resulting effective absorption can be grey if the mean optical depth of clumps is high [(2), (3)]. This phenomenon is not explicitly taken into account in the representation of A_λ, but it would be symptomatic of high values of R_V.

References

Boissé, P. (1990), *A&A*, **228**, p.483
Cardelli, J.A., Clayton, G.C., Mathis, J.S. (1989), *ApJ*, **345**, p.245
Mitskevich, A.S. (1995), *A&A*, **298**, p.219
Voshchinnikov, N.V., Molster, F.J., Thé, P.S. (1996), *A&A*, **312**, p.243

Discussion

Miroshnichenko: What is a level of the errors in A_V you obtained from the modeling of the UV Spectral Energy Distribution for HD 45677?
Moujtahid: The value of A_V depends on the wavelength interval we fit. Generally the relative error on A_V is 16.2%.

SPECTROSCOPY OF B[e] STARS

C. JASCHEK
Observatoire de Strasbourg,
11 rue de l'Université,
67000-Strasbourg, France

When one tries to overview a field like the one of the subject of my talk, one becomes impressed by several facts. In first place the lack of a comprehensive list of objects, in second place by the abundance of literature for some of the objects and the absence of studies of the majority (literature reduced to the description made by the discoverer, dating back 30-50 years) and in third place by the abundance of theories explaining usually one star and generalizing the result to all stars of the ill defined group.

1. List of B[e] stars

As far as point 1 goes, I have started with the list by Mercedes Jaschek and Daniel Egret (1982). At the time of publication it was a comprehensive list, based upon all previous work (like the one by Allen and Swings (1976)) but today it needs an updating. In second place one can use the list by Thé et al (1994). The latter was not done specifically for B[e] stars, but for Herbig Ae/Be stars, so that by principle it needs not to be a complete one. After revising the literature, furnished by the CDS, I collected the stars listed in table 1.

This is my basic list upon which my conference is based. No doubt you will miss some of your favorites. Let me add that I have excluded stars which are composites, like HD 158515 and MWC 623. The reason for doing so is that phenomena in close binaries are sufficiently complex by itself, so that at a first stage it is probably best to set them aside. An asterisk besides the object means that I have seen the spectrum myself, whereas for the remainder the evidence is bibliographic. I should add here the warning that this is definitely not a bibliographic review, so that many authors who have contributed a great deal to our knowledge of the group are seldomly mentioned- this is particularly true for P. Merrill, P. Swings, J. P. Swings and Y. Andrillat. I have quoted just one paper per star in table 1 where a

A. M. Hubert and C. Jaschek (eds.), B[e] Stars, 61–77.

summary of the description of the spectrum is provided.

TABLE 1. List of B[e] stars and bibliography. As explained in the text, the bibliography provides only the reference to a recent paper describing the spectrum.

MWC 17*	Leibowitz 1977, Unpublished
MWC 56*	Swings and Allen 1972, Unpublished
MWC 84*	Downes 1984, Unpublished
MWC 819	Swings and Allen 1973
HD 45677*	de Winter, van den Ancker 1977, Andrillat et al 1997
HD 50138*	Andrillat, Jaschek and Jaschek (in press)
HD 51585*	Jaschek et al 1996a
CD -24 5721	Allen and Swings 1976
Ve 2-27	Allen and Swings 1976
He 2-91	Allen and Swings 1976
AS 269	Allen and Swings 1976
MWC 922	Rudy et al 1992
MWC 297*	Nisini et al 1995, Unpublished
MWC 300	Winkler and Wolf 1989
MWC 314	Miroshnichenko 1996
MWC 342*	Brosch et al 1978
MWC 349*	Andrillat et al 1996
MWC 645*	Jaschek et al 1996b

2. Behavior of the elements

A general definition of B[e] stars is that they are Be stars (i.e. stars with hydrogen line emission) with emission lines of forbidden elements like Fe II. I shall thus proceed by examining first the evidence concerning H, then that concerning the He lines and then the other permitted and forbidden lines existing in these objects.

2.1. HYDROGEN

Figures 1 and 2 display the profiles of some hydrogen lines of the Balmer series observed at the Haute Provence Observatory (OHP) and of some lines of the Paschen series. One is immediately struck by the large variety of profiles and of equivalent widths. Emission is furthermore strongly variable. In some stars it sets in at H7 (MWC 314, Miroshnichenko 1996), in others at H20 (MWC 56, Swings and Struve, 1945).

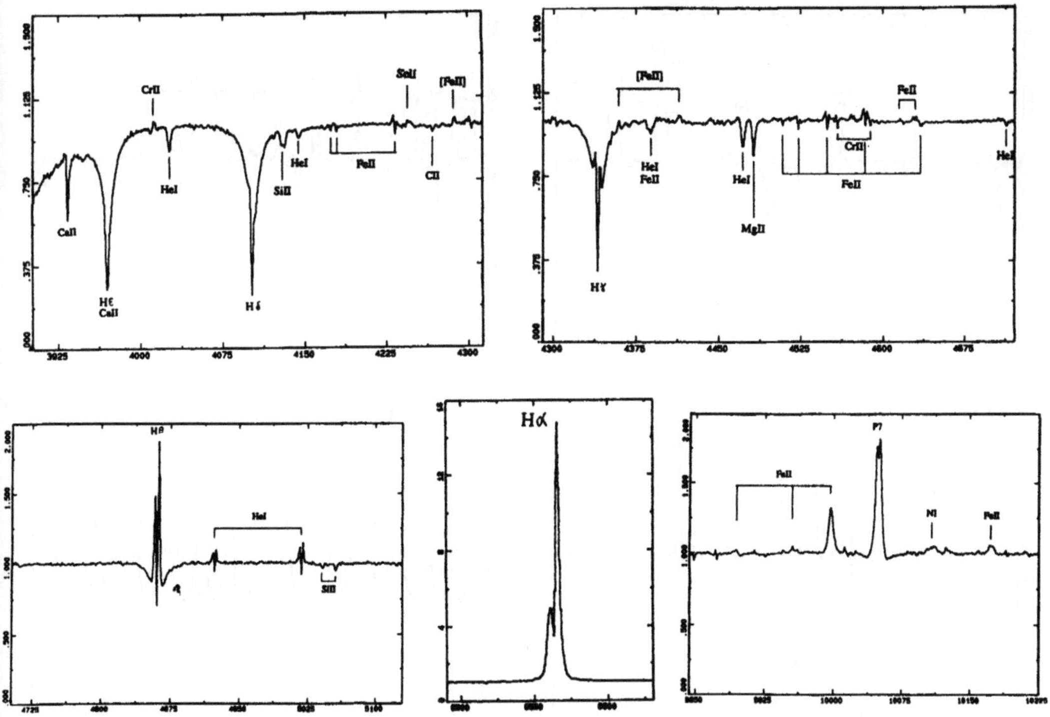

Figure 1. Profiles of some Balmer lines in the spectrum of HD 50138 from Andrillat et al. (1997)

P Cyg profiles may or may not be present (see Figs. 3 and 4). In Paschen one observes the same disparity (see Fig. 5). In table 2, I have collected the equivalent widths of P14 for the stars observed by Mme Andrillat.

Similarly the profiles in the Paschen series may be simple, double, P Cyg type, superposed upon broad absorption lines. If double, the separation of the emission peaks rarely is larger than 200 km/s. In general one sees easier the Paschen emission up to higher terms than the Balmer emission - exactly like in Be stars.

The extreme variability of the hydrogen lines can be seen in figure 3, taken from Israelian et al (1996) for HD 45677. One point which should be stressed is that up to now very few measurements of equivalent widths have been provided - we are still in the qualitative description stage. This applies not only to hydrogen but to all elements. The same is of course true for line profiles.

On the behavior of the other series - Pfund, Brackett - we know very little. One star each has been observed by Nisini et al (1995) (MWC 297) and by McGregor et al (1988) (HD 45677). In both stars Brackett lines were seen in emission and in one also some lines of the Pfund series.

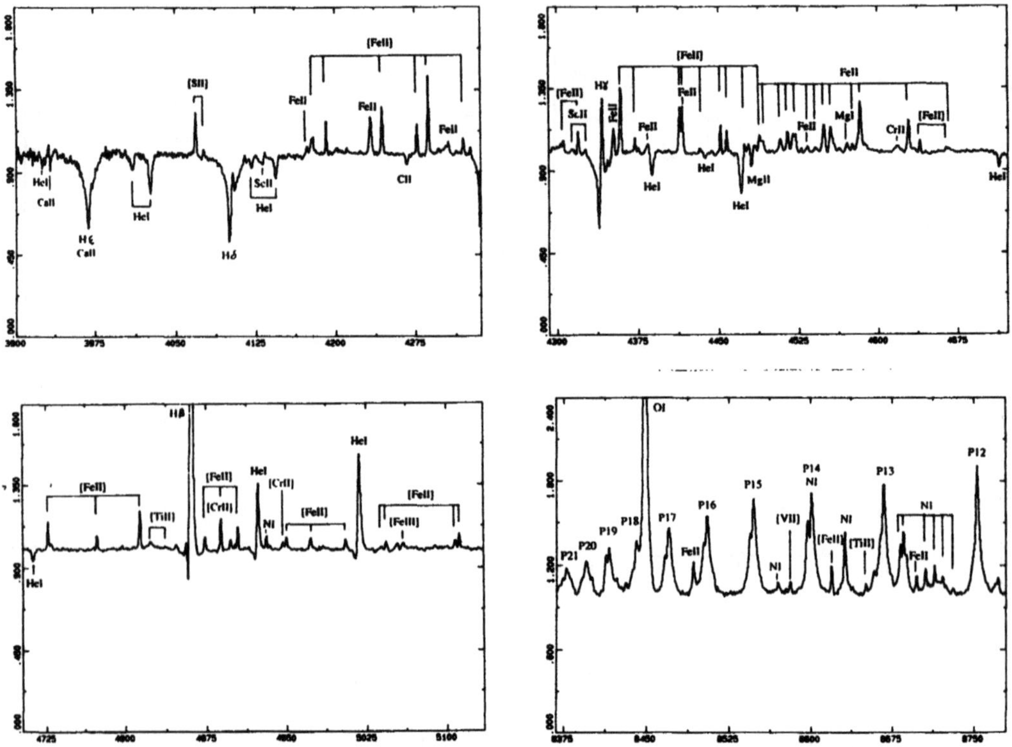

Figure 2. Profiles of some Balmer lines in the spectrum of HD 45677 from Andrillat et al (1997)

TABLE 2. Equivalent widths of P14 (in Å)

MWC 17	25
HD 45677	7.2
MWC 297	in abs.
MWC 56	2.5
HD 51585	6-10
MWC 349	11.4
MWC 84	3.9
HD 50138	1.2-2.2
MWC 645	28.0

2.2. HELIUM

We find again a large variety of situations. In two thirds of the stars of table 1 we find at least some lines in emission; but in five we have no

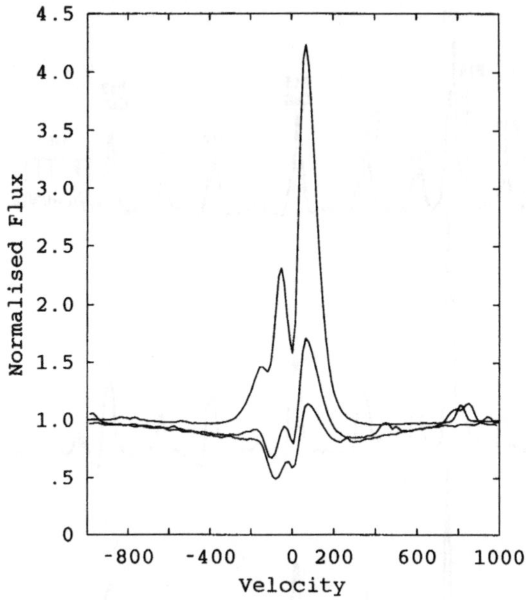

Figure 3. Variations of Balmer lines in HD 45677, from Israelian et al (1996)

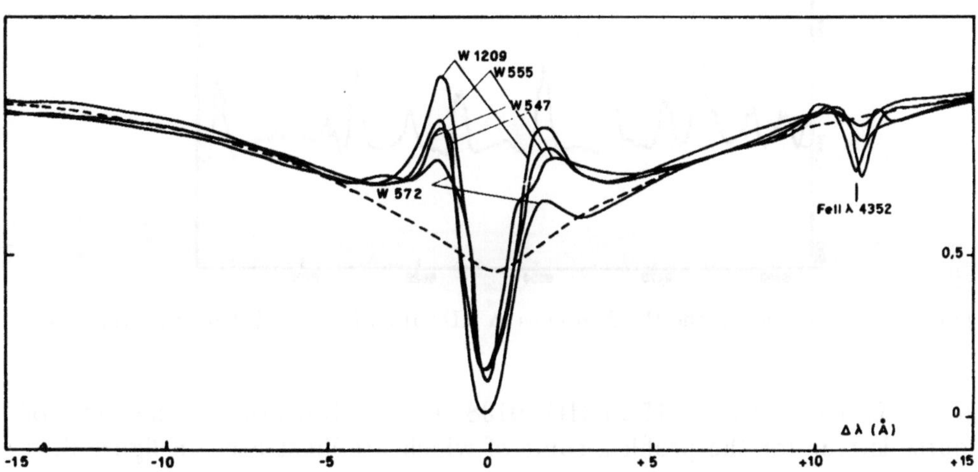

Figure 4. Variations of H5 and Fe II, in HD 50138, from Doazan (1965)

description. On the hand, since the descriptions given in the literature do not cover all the same wavelength range, we are probably safe when saying that in general some He I lines are seen in emission. Here again the variety is very large - λ 6678 has an equivalent width of 40 Å in MWC 481, of 0.93

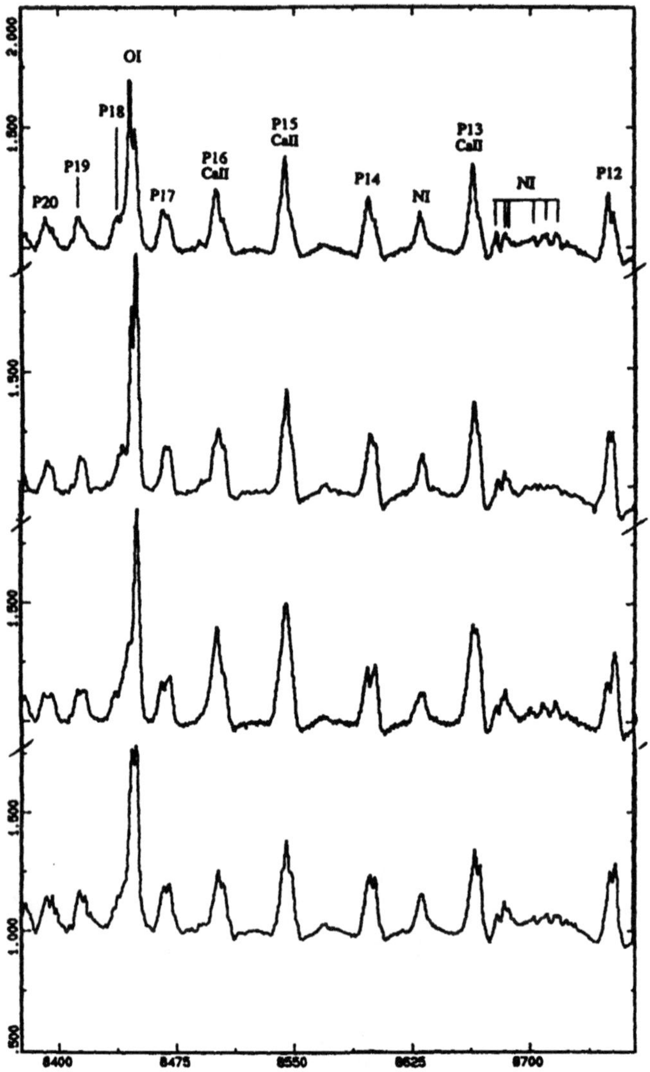

Figure 5. Variations in the Paschen lines in HD 50138, from Andrillat et al. (in press)

in MWC 645 and of -8.15 in HD 50138 (i.e. in absorption). The same of course is true for the profiles, some of which are illustrated in figure 6. If two emission peaks are present, their separation is generally of the order of 200 km/s or more. P Cyg type profiles are a frequent phenomenon, but are certainly not present in all stars.

2.3. LINES OF OTHER ELEMENTS PRESENT

If one has to start with the element that shows the most lines in these stars, this element is Fe. In the stars analysed so far in our series (Y. Andrillat, M.

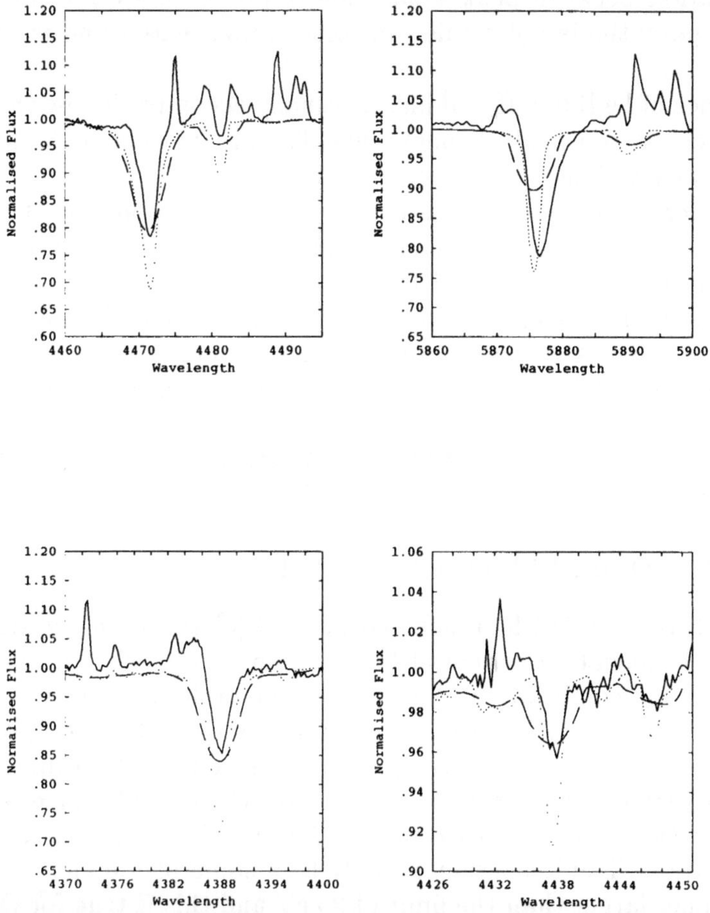

Figure 6. Variations in the profiles of some He I lines, in HD 45677 from Israelian et al. (1996)

Jaschek and C. Jaschek) we find iron lines (both permitted and forbidden) contributing between 53 % (MWC 645) and 33% (HD 50138). This is one of the main characteristics of B[e] stars. In common Be stars Fe lines are neither very conspicuous nor very strong, and [Fe II] lines are absent, as a rule. Fe II lines may be either present in absorption, in emission or in both forms. A common phenomenon is that Fe II lines are formed by an absorption line bordered by two emissions, whose peaks are separated by 50 – 100 km/s. The V/R intensity is variable. The structure is well visible in the most intense lines of the multiplets, suggesting that probably it exists for all lines and it is just a question of resolution. This phenomenon was described in the past for HD 50138 by Doazan (1965). It should be added that a relation between peak separation of double lines and ionization or excitation potential is not a rule - in some stars it exists and in others it is absent.

The presence of Fe II and [Fe II] lines is usually accompanied by the presence of lines of other iron peak elements, like Ti and Cr, whereas V, Mn and Ni are rather rarely found.

Of the lighter elements one finds usually N I in emission and O I. The line 8446 is usually rather strong because of fluorescence with Ly β. But it is also variable - in HD 45677 it varied from 18 to 27 Å in one year. Furthermore a large variety in equivalent width exists between the different stars, which can go from a few Å to several tens of Å. As could be expected, a general relation between W(8446) and W(P14) exists, in the sense that

$$W(8446) \sim 6\, W(P14)$$

2.4. OTHER FORBIDDEN LINES PRESENT

Allen and Swings(1976) had included in the B[e] group only stars in which no ions with ionization potential higher than 25 eV existed, and this is a very reasonable condition. We have listed in table 3 all elements present in the stars of table 1. With regard to the elements of table 3, one should notice that they are not always present - in fact considerable variations exist from year to year. Stars in which "exotic" elements have been found are usually those for which extensive identification lists are available, at high resolution. The ionization potentials of some of the elements present are sometimes larger than the limit of 25 eV and this is true for O II, O III, N II, Ar III, Ne III and Fe III. But it should be added that many of the emission lines of these elements are rather faint, in contrast to elements which are usually present. My personal feeling is that one should not put too much emphasis on the presence of these "exotic" elements. What can be seen easily is [O I], [S II] and [Fe II], and these elements have ionization potentials less than 25 eV.

3. Ultraviolet spectra

Because B[e] stars are usually faint, only a few stars have been observed in the UV, with IUE. In what follows we summarize the little which is known.
HD 45677: Sitko and Savage (1980) describe absorption lines of Si IV and C IV plus emission lines of Fe II, Cr II, He II, C I, N I and Mg II.
HD 50138: Hutsemekers (1985) describes absorption lines of C IV, Si IV,

TABLE 3. Emission lines present in the stars of table 1

MWC 17	[Fe II], [N II], [O I], [O II], [O III], [S II], [Ar III], [Fe III]
MWC 56	no forbidden line emission reported
MWC 84	[Fe II], [S II], [Ti II], [Cr II], [Fe III]
MWC 819	[O I], [Fe II]
HD 45677	[Fe II], [S II], [Ti I], [Ti II], [V II], [Cr II], [Ni II]
HD 50138	[Fe II], [Ti II], [Cr II]
HD 51585	[Fe II], [O I], [O II], [O III], [Ne III], [SII], [Ti II], [V II], [Cr II]
CD -24 5721	[Fe II], [S II]
Ve 2-27	[Fe II], [N II], [O I], [S II], [Fe III], [Ni II]
He 2-91	[Fe II], [O I]
AS 269	[Fe II], [O I], [N II]
MWC 922	[Fe II], [O I], [S II]
CPD -52 9243	[Fe II], [O I]
MWC 300	[Fe II], [O I], [N II], [S II], [Fe III]
MWC 314	[Fe II], [Fe III]
MWC 342	[Fe II], [Fe III]
MWC 349A	[Fe II], [N II], [O I], [O II], [O III], [Ar III], [Ca II], [Ti II], [V II], [Cr II], [Fe III], [Ni II]
MWC 645	[Fe II], [O I], [S II], [V II], [Cr II], [Ni II], [Zr II]

about normal for early type stars, plus lines of Fe III and Al III. He suspects that all lines of doubly ionized elements show a double structure. Of emission lines he finds only Fe II lines and the Mg II doublet.
HD 51585: Houziaux et al (1982) describe the spectrum as dominated by strong emission lines, some of them seen in PN. They mention explicitely O I (1302), [N IV] (1483), C IV (1548), He II (1640), [O III] (1666), [N III], [C III], Fe II , Mg II and S II at 2849.
MWC 314: Miroshnichenko (1996) finds Fe II and Cr II in emission, but Mg II in absorption, but he has only the spectrum from 2500 Å onwards.
In conclusion it can be said that Fe II lines appear here too, plus a list of elements which varies from star to star. As for the absorption spectra they seem to be about normal for early type stars.

4. Spectrum variability

Let us now summarize what we know about variability. We have already mentioned the lack of detailed studies, so that mostly we have qualitative descriptions made by different authors over the years. This suffices in general to show that spectrum variability does exist, but does not teach us very much about the time scales. Without going into too much detail, let

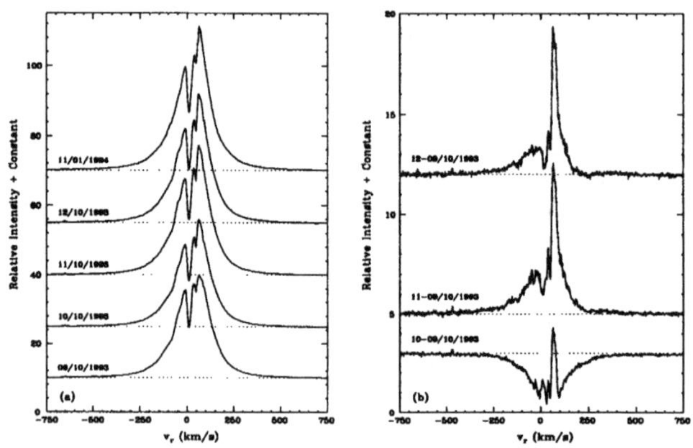

Figure 7. Variations in the profile of Hα, from de Winter and van den Ancker (1997)

us take five objects which are the best studied of the group.
HD 45677: We use for this object the studies by Israelian et al (1996) and de Winter and van den Ancker (1997). From their follow-up of the He I line 5876 (see fig. 7) they conclude the presence of day-to-day changes. Besides this long term variations do exist as evidenced by the photometric behavior. De Winter et van den Ancker (1997) have analysed the photometric and spectroscopic behavior and came to the conclusion that besides the long term variations there exist also intermediate time scale variations and "flickering" with time scales from hours to days. The latter are detected with an analysis of high resolution spectroscopy of Hα, He I and Na I. Based upon our own material, we found changes in W(P14) between 5.8 and 7.2 and of O I 8446 between 18 and 27 in three years.
HD 50138: For this star we have the discussions by Doazan (1965) and Hubert-Delplace and Hubert (1979). Doazan discussed the variations on the basis of high dispersion photographic plates. She found a cycle of about 50 days, present both in the structure and radial velocity of the metallic and the Balmer lines. Hubert and Hubert-Delplace (1979) confirmed the phenomenon of the variable emission structures over period of years.
On our material we found variations of W(P14) between 1.4 and 2.2 Å over the years.

HD 51585: The scanty material does not permit to derive periods. We found on our material many variations over three years; W(P14) between

6.2 and 9.8 Å; of He I 7065 between 8.1 and 13.0 and of Fe II 8490 3.5 and 4.9 to quote but a few values.
MWC 349A: No detailed study exists of the variations, but assembling the existing material it is clear that the equivalent widths of many elements have changed over the years, and this includes forbidden elements like [O III], [N II], [Fe II] and [Fe III]. Generally the changes are of less than a factor two.
MWC 645: Also in this star large variations in equivalent width exist, again of the order of a factor 2. As an illustration of this, we can quote the fact that out of 20 lines identified in an overlapping region ($\lambda\lambda$ 6364-6678) at two different epochs only nine lines do have the same identifications ! (see figs. 9 and 10).
We have illustrated some of the variations found by figure 8 (K lines in HD 50138), figures 9 and 11 (variations in HD 50138). In conclusion it is well established that all these stars - and probably the whole group - shows line intensity variations of at least a factor of two in the equivalent widths. As for time scales there exist certainly year to year and day to day variations, but that is all what one can conclude.

I would like to conclude this section with two remarks. First that although changes exist, there are exceptions - two plates taken at different epochs of the same object may not show changes. Second, that the changes affect all kind of lines, independently of their excitation and/or ionization potential and/or of their character of permitted or forbidden transitions. This implies that all line forming layers of the atmosphere are affected and there exist no abstractions like quiet and disturbed layers.

5. Absorption versus emission line spectra

It is a very obvious phenomenon that the spectra can be divided into two regions- one in which the absorption lines predominate - usually the blue region - and one in which the emission lines predominate. The wavelength at which this happens differs from star to star. One can further have an idea of the temperature of the shell by simply examining which elements are present and with what intensity, using the tables of Jaschek and Jaschek (1995). We have used it to provide an order-of-magnitude estimate of the shell radius, defining by shell radius the one of a sphere which has the temperature of the shell and dominates the black body emission of the underlying star. The method fails of course when no absorption lines are visible.

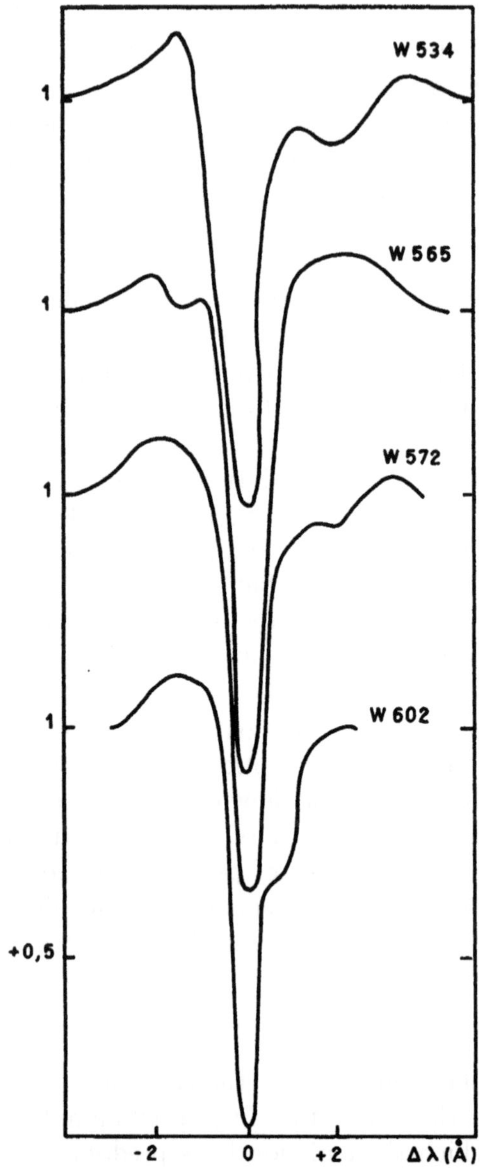

Figure 8. Variations in the profile of the K line in HD 50138, from Doazan (1965)

6. Classification of the underlying spectrum and desiderata for future works

One question which is often asked to classifiers is to provide a spectral type and a luminosity class of these objects. This should be possible in theory, but in practice one finds many pitfalls. In many stars the lines are filled in by emissions, which can only be seen at high dispersion. If not seen the star is easily misclassified. This applies not only to He I but also to C, N and O lines usuable for classification of early type stars. For luminosity classes

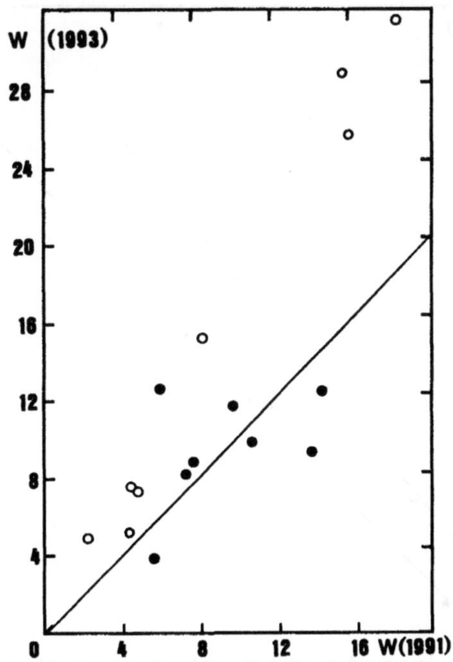

Figure 9. Variations in equivalent width of some lines in MWC 645, from Jaschek et al (1996b)

the problem is even worse and one should not trust them - usually the only thing one can say is that they are main sequence, supergiants or somewhat in between. I know that this does not help very much, but this is, sorrily, the situation.

For instance for HD 50138 Hδ corresponds to B5, He I to B7 III, Ca II to A0, Si II to B5 III, Mg II to B5 Ia, and C II to B5 III. We called the object B5 (III). Further inconsistencies turn up in the few stars in which we have UV spectra- as a rule the UV types do not agree with the types derived from the 4000 Å region. In HD 50138 the UV type is B9 pec.

I have collected in table 4 what seem to be available in matters of classification.

You will probably agree that on the whole the situation is unsatisfactory. It could be a nice exercice to turn now back to table 1 and eliminate all those objects which do not satisfy the behavior of the majority, and then to select one, the so called typical object in order that we can make a satisfactory theory for it and apply it to all other stars of the group. One should however keep in mind that before reaching such a happy stage it would be more profitable to try to observe a little bit more, in particular to solve a number of open questions. I list a few, but the list can be enlarged at you

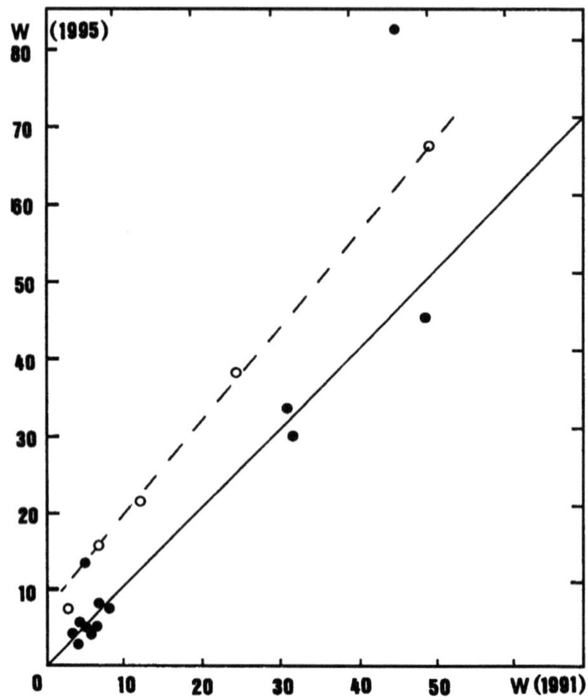

Figure 10. Variations in equivalent width of some lines in MWC 645, from Jaschek et al (1996b)

TABLE 4. Spectral classifications of B[e] stars

MWC 56	This object has a late G III companion
MWC 819	no absorption lines visible.
HD 45677	B2 (IV)
HD 50138	B5 (III)
HD 51585	no absorption lines visible.
MWC 922	B3-6 ?
MWC 297	type B0 (Nisini et al 1995)
MWC 300	B1 Ia+ (Winkler and Wolf 1989)
MWC 314	B0 Ia (Miroshnichenko 1996), B2 (Swensson 1943)
MWC 342	B3 III (Brosch 1978)
MWC 349A	no absorption lines visible. Shell B-type.
MWC 645	no absorption lines visible. Shell late B-type.

wish.

1- to obtain spectra over as long a wavelength interval as possible and

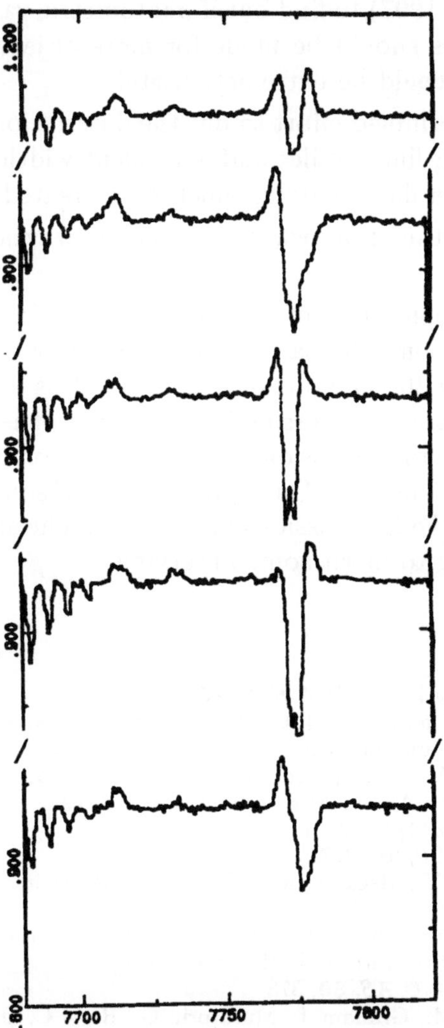

Figure 11. Variations of the profile of O I l 7776 in HD 50138, from Andrillat et al (in press)

with the highest feasible resolution, in order to carry out satisfactory identifications. If this could be carried out for all stars of table 1, we could at least have a larger number of recognized B[e] stars and would not be obliged to work always on the five or six primadonnas of the group.

2 - to concentrate efforts on a few objects for documenting the time variations. Up to now we have no clear idea if hourly, daily, weekly, monthly or yearly changes predominate. Or is the activity dominated by outbursts like the ones documented by Hutsemekers for HD 50138 ? Efforts like the one made by Swings et al (1980) for HD 45677 and repeated by de Winter

and van den Ancker (1997) to find out if photometric variations go with the spectroscopic changes should be made for more objects. An international observing campaign could be extremely useful.

3 - to make a determined effort to put the descriptions on a quantitative footing, by publishing line profiles and equivalent widths. Such information can be stored easily in data centers, which were created for such a purpose.

4 - to examine other practically unexplored wavelength regions (UV, $\lambda \geq 1\mu$).

5 - to follow the suggestion of the late Mercedes Jaschek that B[e] stars are closely related to some Be stars, namely those of what we called group I. These are early type (B0-B3) stars of luminosity class V to III which exhibit permanent Fe II emissions. These emission vary in phase with Balmer lines and present often double emission structures. In some of these stars one observes also emission in He I. On reproductions of some spectra it seemed to me that traces of [Fe II] emissions are present. The suggestion is anyway sufficiently promising to merit some observing.

References

Allen, D.A. and Swings J.P. (1976) *A & A*, **47**, 293
Andrillat Y., Jaschek M. and Jaschek C. (1996) *A & AS*, **118**, 495
Andrillat Y., Jaschek C. and Jaschek M. (1997) *A & AS*, **124**, 441
Brosch N., Leibowitz E.M., and Spector N. (1978) *A & A*, **65**, 259
de Winter D. and van den Ancker M.E. (1997) *A & AS*, **121**, 275
Doazan V. (1965) *Ann. d'Aph.*, **28**, 1
Downes R.A. (1984) *PASP*, **96**, 807
Houziaux L., Andrillat Y., Heck A. and Nandy K. (1982) in *IAU symp.* **98**, eds M. Jaschek and H.J. Groth, p. 427
Hubert-Delplace A.M. and Hubert H. (1979) *Un atlas des étoiles Be*, Observatoire de Paris-Meudon and Observatoire de Haute-Provence
Hutsemekers D. (1985) *A & AS*, **60**, 373
Israelian G., Friedjung M., Graham J., Muratorio G., Rossi C. and de Winter D. (1996) *A & A*, **311**, 643
Jaschek M. and Egret D. (1982) *Publ.Spéc. CDS*, **4**
Jaschek C. and Jaschek M. (1995) *The behavior of chemical elements in stars*, Cambridge Univ. Press
Jaschek C., Andrillat Y. and Jaschek M. (1996a) *A & AS*, **117**, 281
Jaschek M., Andrillat A. and Jaschek C. (1996b) *A & AS*, **120**, 99
Jaschek C. and Andrillat Y. (1997) *A & A*, in press
Leibowitz E.M. (1977) *A &A*, **56**, 265
McGregor P.J., Hyland A.R. and Hiller D.J. (1988) *ApJ*, **324**, 1071
Miroshichenko A. (1996) *A & A*, **312**, 941
Nisini B., Milillo A., Saraceno P. and Vitali F. (1995) *A & A*, **302**, 169
Rudy R.J., Erwin P., Rossano G.S. and Puetter R.C. (1992) *ApJ*, **398**, 278
Sitko M.L. and Savage B.D. (1980) *ApJ*, **237**, 82
Swensson J.W. (1942) *ApJ*, **97**, 226
Swings P. and Struve O. (1945) *ApJ*, **101**, 228
Swings J.P. and Allen D.A. (1972) *PASP*, **84**, 573
Swings J.P. and Allen D.A. 1973) *Ap.L.*, **14**, 65

Swings J.P., Barbier R., Klutz M., Surdej A. and Surdej J. (1980) *A & A*, **90**,116
Thé P.S., de Winter D. and Pérez M.R. (1994) *A & AS*, **104**, 315
Winkler H. and Wolf B. (1989) *A & A*, **219**, 151

Discussion

Viotti: Concerning the different spectral classifications of the same star according to different absorption lines, this should be related to the different depth of line formation in the very extended atmosphere - resonance lines being for instance formed in the outer lower density regions. It would be important to concentrate our study on lines formed deep in the atmosphere and in spectral regions where the continuum opacity is low, such as in the space UV- In any case, to simplify the problem of classification of B[e] stars, one solution could be to take a star as the best representative of the category - and this star should be HD 45677 - and compare other stars with it.
Jaschek: I agree with your first comment. As far as selecting one typical star I would prefer to wait until we have observed more B[e] stars.
Hubert: One star of your list has Fe II lines in absorption also and what about [Fe II] lines. In classical Be stars there is a general trend: strong hydrogen emission associated with Fe II line emission plus infrared excess.
Jaschek: The star with no [Fe II] is MWC 56, but it had these lines in the past. So in both cases we are dealing with variability- we just catched the star at an inactive phase.
Cassinelli: The difficulty you have in deriving spectral classification reminds me of the problem theorists had in specifying the effective temperature of stars with extended stellar atmospheres. The problem with extended atmospheres is that there is no well defined stellar radius. Instead the radius (say at $\tau_\lambda = 1$) is wavelength dependent; R_λ may be large at IR wavelengths and small at UV wavelengths. Given that this problem exists for the B[e] stars, observers must not only consider the behaviour of a line (or lines) of a given chemical element. They must also note the wavelength and the likely continuum formation radius at that wavelength. Of course this is difficult without having detailed extended model atmospheres. But, in general, I would expect that lines in the near IR region be formed far out in the atmosphere, lines in the near UV are formed deep in the atmosphere, where the radiation temperature will be higher.
Jaschek: Yes, I agree with the difficulties you mention. I thought simply (as other observers did too) that when we see the absorption lines, we see the undisturbed layers of the underlying star. In a restricted wavelength region ($\lambda\lambda$ 4200-4500) this should converge to one classification, but it does not, even in such a small region.

THE SELF ABSORPTION CURVE METHOD FOR THE ANALYSIS OF EMISSION LINES OF COMPLEX IONS AS A DIAGNOSTIC OF THE STRUCTURE OF STELLAR ENVELOPES

M. FRIEDJUNG
Institut d'Astrophysique (CNRS), 98 bis Boulevard Arago, 75014 Paris, France

Abstract. A semi-empirical method has been developed to study the emission lines of ions with complex spectra such as Fe II and Ti II, formed in many sorts of stellar envelope. Information can be obtained notably about the excitation conditions, the column density and the dimensions of the line emitting region, without making detailed model dependent complex radiative transfer computations. This method is particularly useful in the study of B[e] star envelopes.

1. Introduction

Metallic emission lines from ions such as Fe II, Cr II and Ti II (and more rarely Fe I) are seen in the spectra of many classes of star. These lines are generally not emitted in normal stellar photospheres, but rather in other less well understood regions further out, which can include winds, "chromospheres" and disks of various kinds. The Grotrian diagrams of these ions are very complex and when many lines are optically thick, theoretical radiative transfer calculations though possible with present day computers, become quite model dependent. This is embarassing when the general nature of the line emission region is badly known. A simple semi-empirical less model dependent approach has been developed, which can give information about line emission regions, such as those of B[e] stars. The work to be described has been done in collaboration with a number of people, including G. Muratorio, C. Rossi, R. Viotti and G.B. Baratta.

A. M. Hubert and C. Jaschek (eds.), B[e] Stars, 79–83.

2. Basic principles of the method

Let us to start, consider a very simple situation, where a plane parallel slab emits in a spectrum line with a rectangular profile. The slab is homogeneous and perpendicular to the line of sight of an observer, whose distance from the slab is much greater than the extent of the slab perpendicular to the line of sight. In this case it is very easy to show, integrating over different optical depths, that:

$$F_{ul} = \frac{gf}{\lambda^3}\frac{2\pi e^2 h}{m_e}\frac{S}{d^2}\frac{N_u}{g_u}\frac{1-e^{-\tau_o}}{\tau_o} \qquad (1)$$

F_{ul} is the total flux of the line, m_e the electron mass, N_u the upper level column density, S the surface area of the slab and d its distance. The total optical thickness of the slab is equal to:

$$\tau_o = gf\lambda\frac{N_l}{g_l}\frac{\pi e^2}{m_e c}\frac{1}{v_o} \qquad (2)$$

Here N_l is the lower level column density and v_o is the width of the line in velocity units. In this discussion N_l/N_u is assumed large, so the effect of stimulated emission is negligible.

In general an emission line will be formed in an inhomogeneous region, while the profile will not be rectangular. Taking logarithms, we replace $log\frac{1-e^{-\tau_o}}{\tau_o}$ in the logarithm of eq. (1) by a general function $Q(\tau_o)$, which gives the effect of self absorption of the line. Then eq. (1) is replaced by:

$$log\,\frac{F_{ul}\lambda^3}{gf} = log\,\frac{2\pi e^2 h}{m_e} + log\,\frac{S}{d^2} + log\,\frac{N_u}{g_u} + Q(\tau_o) \qquad (3)$$

A graph of $Q(\tau_o)$ against the logarithm of τ_o, a characteristic optical thickness of the line emitting region in a certain direction, is a curve, which we call the "self absorption curve" or "SAC". The left hand side of such a curve for optically thin lines is flat and horizontal; the curve bends downwards as τ_o increases. As long as the values of N/g (both for upper and lower levels) are the same inside the same spectroscopic term, a graph of $log\frac{F_{ul}\lambda^3}{gf}$ against $log\;gf\lambda$ for lines of the same multiplet will have points lying on an observational self absorption curve, as can be seen from eqs. (2) and (3). The only thing which then varies in the abscissa is $gf\lambda$, proportional to τ_o. The curves for other multiplets with different values of N_l/g_l and N_u/g_u will be shifted. Multiplets having the same lower term (same N_l/g_l) will have curves vertically shifted with respect to each other, while those having the same upper term (same N_u/g_u) will have the curves shifted horizontally with respect to each other. In addition, making a second assumption, as

long as the shape of the SAC is the same for all multiplets (true at least in the simpler theoretical situations), one can simultaneously find the relative populations of the upper and lower terms using the SAC method.
The SAC method is much better than the old emision line curve of growth method. The curve of growth for a particular multiplet is a graph of $log\ F_{ul}$ against $log\ gf\lambda$. It is not possible to simultaneously determine the relative populations of the upper and lower terms, by comparing the emission line curves of growth of different multiplets. Only excitation temperatures, assuming one Boltzmann law for the level populations, can be found.
Taking eqs. (2) and (3) and subtracting the logarithm of eq.(2) from eq. (3) at $\tau_o = 1$, we obtain:

$$log(\frac{F_{ul}\lambda^3}{gf})_{\tau_o=1} = Q(1) - log\ gf\lambda + log\frac{S}{d^2} + (log\frac{N_u}{g_u} - log\frac{N_l}{g_l}) + log\ v_o + log\ 2hc \quad (4)$$

This can be used to determine S if information is available about the ratio of upper to lower term populations and about the value of v_o. Similarly eq. (2) can be used to obtain lower term column densities.

3. Application of the method

Observational SACs were determined by Friedjung and Muratorio (1980) for the B[e] star S22. $log\ \frac{N_u}{g_u} - log\ \frac{N_l}{g_l}$ was estimated by comparing permitted and forbidden lines, because the lower levels of permitted Fe II lines are the upper levels of [Fe II] lines. Assuming the formation of both in the same region gave an upper limit to S (equal to πR_o^2, with R_o a characteristic radius of the line emitting region). Theoretical Fe II SACs were calculated by Friedjung and Muratorio (1987) for lines formed in a stellar wind, assuming Viotti's (1976) optically thin line excitation theory with different hypotheses and applied to B[e] stars by Muratorio and Friedjung (1988). Populations appeared to be consistent with approximately the same Boltzmann laws being true for the upper and lower levels. Evidence was obtained supporting the idea of the formation of Fe II emission lines in circumstellar disks rather than in spherically symmetric winds. The method was applied by Viotti et al (1991) to the LBV star AG Car.
The SAC method is however often hard to apply in practice because in many cases the observations show no unambiguous sign of SAC curvature and are consistent with its being linear. In more recent work we have either assumed a linear SAC when indicated by the observations or a schematic mathematical formula for the curve with several parameters. Independent Boltzmann laws are used for the upper and lower levels, deviations from such an assumption being detectable. The linear SAC approach was first

used for the VV Cephei type binary KQ Pup by Muratorio et al (1992). It has been improved in the study of the Herbig Ae/Be star V380 Ori (Muratorio et al 1997, Rossi et al 1997). We use the 2 Boltzmann laws to superpose the multiplet SACs at 0 ev. The composite SAC is then a graph of Y against X, where X and Y are given by:

$$X = log\ gf\lambda - p\chi_l + r + log\ \frac{\pi e^2}{m_e c} \tag{5}$$

$$Y = log\ F_u l \frac{\lambda^3}{gf} + q\chi_u \tag{6}$$

Here χ_u and χ_l are the upper and lower level excitation potentials of each spectrum line, while p, r and q are constants to be determined by a least squares solution for minimum scatter of the graph. The use of two different Boltzmann laws is justified for the strong optical and ultraviolet lines of Fe^+ as the lower terms belong to even configurations up to around 4.5 ev, while the upper levels are those of odd parity configurations of a higher excitation potential. Ti II and Cr II have a similar behaviour.

Let it also be noted that when the SAC is linear, it is only possible using eqs. (2) and (5), to find a minimum τ_o at zero electron volts for the lines with the smallest values of X. In this way only minimum column densities will be obtained.

Future work on the SAC method should among other things include a theoretical study of what are the conditions for its basic assumptions to be valid. For this radiative transfer calculations need to be performed for different situations.

Finally the author would like to thank Roberto Viotti and Gerard Muratorio for reading and suggesting improvements to this text.

References

Friedjung, M., Muratorio, G. 1980, A&A **85**, 233

Friedjung, M., Muratorio, G. 1987, A&A **188**, 100

Muratorio, G., Friedjung, M. 1988, A&A **190**, 103

Muratorio, G., Viotti, R., Friedjung, M., Baratta, G.B., Rossi, C. 1992 **258**, 423

Muratorio, G., Rossi, C., Friedjung, M. Viotti, R. 1997, proceedings of this conference

Rossi, C., Errico, L., Friedjung, M., Giovannelli, F., Muratorio, G., Viotti, R., Vittone, A. 1997, A&A, submitted

Viotti, R. 1976, ApJ **204**, 293

Viotti, R., Baratta, G.B., Rossi, C., di Fazio, A. 1991, Symp IAU 143, "Wolf Rayet Stars and Interrelations with other Massive Stars in Galaxies ", (eds) K. A. Van der Hucht and B. Hidayat, Kluwer, p 499

Discussion

Zickgraf: We have recently studied the UV line profiles of a few B[e] supergiants including Hen S 22 and R 82. We found that the Fe II resonance lines are in fact optically thick. In the case of Hen S 22 the Fe II lines exhibit P Cyg profiles.
Friedjung: The UV resonance lines must be optically extremely thick, because of the already very large optical thickness of the optical lines. The P Cyg absorption components in the ultraviolet of Hen S 22 could be due to a wind, other than that in the disk, whose P Cyg absorption components are too weak in the optical to be seen here.
Viotti: I would like to remark that also the forbidden lines might be optically thick. This is in fact what we have found for the [Fe II] lines in LMC star Hen S 22.

THE HIGH EXCITATION FE II LINES AS A DIAGNOSTICS OF THE OUTER ENVELOPES OF EARLY TYPE EMISSION LINE STARS*

R. VIOTTI[1], A. DAMINELI[2], C. ROSSI[3]

1. Istituto di Astrofisica Spaziale, CNR, Frascati, Italy
2. JILA, Boulder, CO, USA, and IAGUSP, São Paulo, Brazil
3. Istituto Astronomico, Università La Sapienza, Roma
(*) Based on observations collected at OHP (France) and LNA (Brazil).

Abstract. The analysis of the spectrum of emission line stars in a wide spectral range has put in evidence a number of high excitation emission lines having anomalously high intensities. We illustrate the case of the high excitation Fe II lines observed in the ultraviolet and near–IR of different stellar categories, and discuss the information which can be derived from their study.

1. Observational aspects

This work is deadling with a rather hard topic, and it is intended to stimulate researchers' attention to the problem of *selective line excitation*, which is not only limited to the well known classical cases of the Bowen mechanism for the near-UV lines of O III, but it is a more general feature especially in B–type emission line stars. As for example, Hamann et al. (1994) described the red spectrum of η Car observed at high resolution, and identified a great deal of emission lines mostly belonging to transition from highly excited levels of Fe^+. Particularly intense is the Fe II 849 nm line near P16, which Hamann et al. attributed to cascade from excited states pumped by resonant absorption of Lyα. Moving to shorter wavelengths we find in the UV spectrum of this star the highly excited Fe II lines at 178 nm (UV multiplet 191) and at 251 nm (Kurucz 1981, not in Moore's UV Multiplet Tables) are very strong in emission. The latter actually is the most

A. M. Hubert and C. Jaschek (eds.), B[e] Stars, 85–89.

prominent emission feature in the low resolution IUE spectrum of this star (Viotti et al. 1989).

Moving to the near–IR, an interesting case is represented by the classical Be star γ Cas which is showing at 999.7 nm a prominent emission of the high excitation Fe II line, listed in Kurucz' (1981) tables, which has an intensity close to that of the nearby hydrogen line Pδ (Viotti et al. 1997, see Figure 1). Also the profiles of the two lines are quite similar in agreement with Johansson (1984) suggestion that the anomalous strength of the 999.7 nm Fe II line could be due to selective excitation by fluorescence from Lyα. In this case the observation of this line at high resolution can supply additional information on the H II circumstellar region.

A systematic study of this region (Torres 1994) has revealed that a large number of B emission stars have a prominent 999.7 nm line. We have also identified this line weakly in emission in the "normal" late B–type supergiant β Ori (Fig.1).

Variability is a common feature of B–emission stars, which can be used as a powerful mean for understanding the physical processes of line formation. For instance in the case of η Car the 251 nm feature disappeared during the 1980-81 shell episode of the star (Damineli et al. 1995), as a consequence of the decrease of the emitting envelope excitation. The dramatic spectral evolution of the LBV star AG Car during its last brightening is well illustrated by the variation of the Fe II 999.7 nm and Pδ lines (Figure 1). The cooling of the star's spectrum is marked by the increase of the two lines, while variation of the Fe II/Pδ ratio is most probably an effect of the larger opacity of the hydrogen emission line.

2. Fe II level population

We show Fig.2 the schematic level population of Fe^+. In the case of line formation outside LTE, we expect large departures of the level population from a Boltzmann–type distribution law with a single excitation temperature T: $log N/g \sim exp(-\chi/kT)$. Generally, at electron densities larger than about 10^6 cm^{-3} the lower metastable levels are in statistical equilibrium, while the levels above about 4.7 eV are populated by continuum fluorescence from the ground and metastable levels from the diluted radiation of the underlying star (Viotti 1976). We also expect a large contribution of the *line fluorescence* due to the accidental coincidence of the energy of the upper level of the excited line with the photon energy of a strong line of another species such as the field Lyα radiation. This is the same as the Bowen mechanism which, in the case of Fe^+, is applied to a large number of transitions. *Dielectronic recombination* which is populating the upper levels of the recombined species is a process to be considered especially in

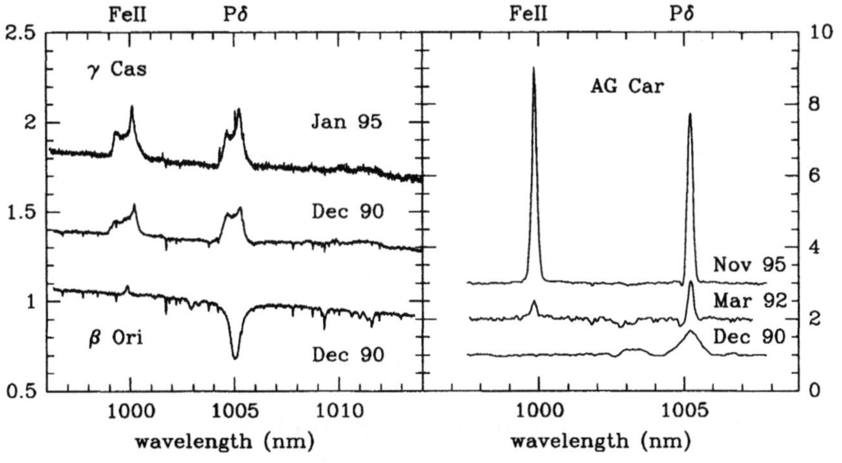

Figure 1. The near–IR spectrum of B–emission stars near the high excitation Fe II 999.7 nm line. *Left.* OHP–AURELIE observations of γ Cas and β Ori. *Right.* LNA (Brazil) observation of the spectral variability of AG Car.

complex ions, though its effects are very hard to compute. It is anyhow the dominant process of ionization equilibrium (balanced by autoionization) in complex atoms.

As discussed in this Volume by M. Friedjung, a statistical analysis such as the SAC, is needed to study the complex mechanisms of level population, taking into account radiative transfer problems in a medium which frequently appears to be optically thick also to the weaker emission lines. The M2Iab+B0V[e] binary KQ Pup displays a rich Fe II emission line spectrum formed in the cool star wind heated by the B–star radiation. Very strong in emission are the lines of the 178 nm and 251 nm multiplets. The SAC analysis of the UV Fe II emissions (Muratorio et al. 1992) has disclosed a large number of lines whose intensity is larger or much larger than that expected if the upper levels were populated according to a Boltzmann–type law (Fig.2). Also the scatter of the individual points is larger than expected from the observational errors. We therefore conclude that some levels, if not all, are populated by fluorescence excitation (mainly Lα) and/or by dielectronic recombination.

In conclusion, for the Fe II emission lines (and, most probably, for any other complex ion as well), selective mechanisms (fluorescence, possibly dielectronic recombination) are the main contributors to the level population.

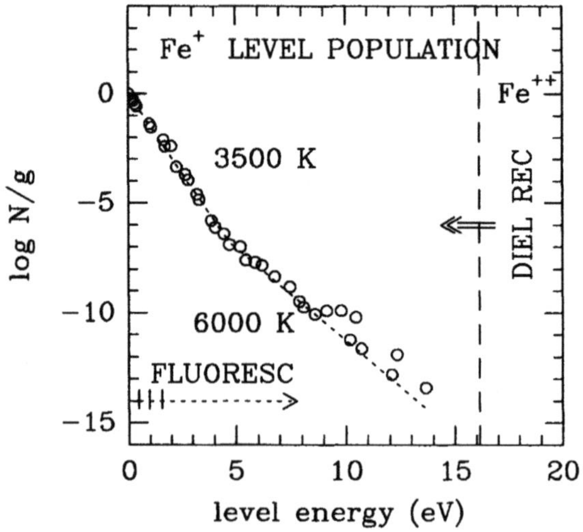

Figure 2. Schematic representation of the Fe^{+} level population in the case of line formation outside LTE. Note the different mean excitation temperature of the lower and upper levels.

Hence, any analysis of their intensity should first addressed to the determination of the level population, such as with the SAC method, which is a way to select among the different excitation processes. For this purpose, the study of the spectral variations would also be of the greatest help.

We are grateful to M. Friedjung for discussions. A.D. was partially supported by FAPESP.

References

Damineli A., Cassatella A., Viotti R., et al. (1995), *Rev. Mex. A. Ap.*, **2**, 41
Hamann F., DePoy D.L., Johansson S., Elias J. (1994), *ApJ*, **422**, 626
Kurucz R.L. (1981), *Smithonian Astrophys. Obs. Special Report*, **390**
Johansson S. (1984) *Phys. Scripta*, **T8**, 63
Muratorio G., Viotti R., Friedjung M., et al. (1992), *A&A*, **258**, 423
Torres R.M. (1994), *Master Thesis*, IAGUSP, São Paulo, Brazil
Viotti R. (1976), *ApJ*, **204**, 293
Viotti R., Rossi L., Cassatella A., et al. (1989), *ApJS*, **71**, 983
Viotti R., Rossi C., Muratorio G. (1997), *A&AS* in press

Discussion

Hubert: What will high spectral resolution imply for your method?

Viotti: High resolution is needed to have information on the velocity field in the "light" of different lines. Since the velocities involved are of the order of 10^1 to 10^2 km.s^{-1}, a resolution $\geq$ 10 000 is required.

NITROGEN V IN THE WIND OF THE PRE-MAIN SEQUENCE HERBIG Ae STAR AB AUR

J.-C. BOURET AND C. CATALA
Laboratoire d'Astrophysique de Toulouse, Observatoire Midi-Pyrénées, 14 avenue Edouard Belin, 31400, Toulouse, France

AND

T. SIMON
Institute for Astronomy, University of Hawaii, 2680 Woodlawn Drive, Honolulu, Hawaii 96822, USA

Abstract. Recent GHRS observations of the Herbig Ae star AB Aur demonstrate the presence of N V in the wind of this star. We found that models including hot (T≈140,000 K) clumps are able to reproduce the available observations and are also consistent with its X-ray emission.

AB Aur was observed at intermediate spectral resolution with the Goddard High Resolution Spectrograph aboard the Hubble Space Telescope on 1996 February 23, in the spectral region between 1226–1253 Å which surrounds the N V doublet at 1238.8, 1242.8 Å. We identified the emission feature detected at $\lambda \sim 1238.8$ Å as one component of the 2s-2p resonance doublet of N V. Although the other component of this doublet is totally blended with an absorption component, a very weak emission seen at its wavelength is probably due to this second line of the doublet.

The presence of N V resonance lines lead us to introduce an additional high temperature zone to the current model of AB Aur's wind (Catala & Kunasz 1987). Spherically symmetric models produce features which are not observed, but that could be made undetectable by lowering the filling factor of the hot zone. We reduced this filling factor by assuming the presence of corotating interacting regions generated by a fast and slow stream structure controlled by the surface magnetic field. This assumption is supported because azimuthal structures are known to exist in the wind, as witnessed by the rotational modulation detected in some lines of AB Aur; The temperature in the CIRs being expected to rise far above that of the

A. M. Hubert and C. Jaschek (eds.), B[e] Stars, 91–95.

unperturbed stellar wind, we assumed that the N V resonance lines can originate in part of the CIRs, while X-ray emission is produced in the hottest parts of them.

For the solution of the transfer equation, we have used the method of the equivalent two-level atom in the co-moving frame. The presence of shocks, producing hot clumps of material in the wind, was simulated by a two-step procedure. First, we built a spherically symmetric model including a geometrically thin shell of high temperature, with a very steep temperature gradient at its edges. The radiative transfer and statistical equilibrium equations were solved simultaneously for this model, yielding the source functions. Second, we calculated the emergent intensity on a set of rays parallel to the line of sight from the observer, in the usual manner. The emergent flux was then computed by integrating the emergent intensity on the different rays, but we introduced the departures from spherical symmetry by weighting this integration by a function $r(\theta, \phi)$, representing the distribution of shocks in the wind. The assumption behind this procedure is that N V is present only in the hot clumps produced by the shocks, so that its opacity is non-zero only on rays that intercept hot clumps.

The observed N V line is fairly well reproduced by such a model, assuming a filling factor of 7×10^{-3} for the CIR, and temperatures of the order of 140,000 K. If we assume that higher temperatures, with range 1-2 $\times 10^6$ K are also present in other regions within the CIRs, then the same filling factor is consistent with the observed X-ray flux of AB Aur.

When evaluating the amount of non-radiative energy deposited in the wind, we have found that the presence of CIRs keeps the global energetic balance of the wind of AB Aur almost unchanged. Other processes than dissipation of kinetic energy in the CIRs must be present to account for these losses. If present a magnetic field of intensity about 100G would constitute a comfortable reservoir of energy to account for the observed phenomena.

These results provide clues that the wind of AB Aur is very inhomogeneous and includes azimuthal structures giving rise to corotating interaction regions. All the features produced in the CIR, like the N V lines or the X-ray flux of AB Aur as observed by ROSAT are expected to be variable, so it would be of utmost interest to study their variability to obtain valuable informations on the structure of AB Aur's wind.

References

Catala C., Kunasz P.B., 1987, A&A, 174, 158
Catala C., Praderie F., Felenbok P., 1987, A&A, 182, 115
Mihalas D., Kunasz P., 1978, ApJ, 219, 635
Mullan D., 1984, ApJ, 283, 303
Zinnecker H., Preibisch Th., 1994 A&A 292, 152

Discussion

Cassinelli: I wonder whether the CIR model could in fact provide the X-rays that are seen, because one would require a relatively large collision speed. I would think that a better explanation of the X-rays would be magnetic loops at the base of the wind. The N V would thus be formed either in the loop region or by the Auger effect operating in the wind.
Bouret: Indeed it seems that the collision speed we used, which is deduced from observations is not high enough to explain the X-ray luminosity of AB Aur. However we have no informations on the true nature of the shocks; if they are MHD shocks, as in the solar case for example, we could then have enough energy to explain the X-ray emission. Whatever the energy source is, we need to dissipate it into the wind. We can assume two different scenarios: either the energy dissipated in the wind heats some region to about 140 000 K as we found and the N V is generated in such regions, or the energy heats the wind up to 10^6 K and the X-ray flux produced then is used to form N V further in the wind. We did not investigate this second mechanism because we have considered it more complicated than the first one. Moreover several observations show a rotational modulation for different lines, which is easily interpreted in terms of CIR's.
The energy dissipated in reconnexion mechanism would be of the same order of that available in MHD shocks we have invoked so as to explain the measured X-ray flux. So as to discriminate between the two mechanisms, it is necessary to make further observations, to study the variability of the N V lines and of the X-ray flux. If a variability is observed, then the CIR's hypothesis is confirmed or at least strengthened; if no variability is seen, reconnection mechanism "overtakes" the first one.
Langer: You said that the high energy radiative losses can not be compensated by the kinetic energy of the wind flow, but rather that you need a magnetic field. What, in this case, is finally the source of the energy?
Bouret: May be I was not clear enough. I just said that the total kinetic energy of the wind is not large enough to explain the radiative losses we have calculated. Then other processes of energy dissipation must be present. It can either be of magnetic origin, or something completely different. However we have now several observations which strenghten the idea that the magnetic field exists in the wind of AB Aur. The source of energy needed to generate the magnetic field could be found in shear mechanism between two layers with different rotational velocities, under the surface of the star (Lignières, Catala, Mangeney A&A 1996, 314, 465).
Bjorkman: What did you assume for the distribution/geometry of the shocks that produce the N V profiles? When calculating your synthetic profiles what did you assume for the inclination angle of the system?

Bouret: We assumed that the shocks occur where fast streams overtake the slow ones. We have derived constraints on the location of the CIR's by imposing that all lines that had been previously modelled with the "current" model of AB Aurs wind must be kept unchanged by the presence of the CIR's; therefore the latter can only range in a very narrow region within the wind. Moreover we have used the X-ray luminosity measured by ROSAT to gain informations on the size, location and filling factor of the high temperature zone which is used to simulate the presence of shocks in the wind. Of course the distribution of shocks in the wind we found, is likely to be not quite right: from the synthetic profile we can deduce that it needs either more shocks at zero velocity, or else a higher circular velocity (assumed to be 100 km.s^{-1}) to spread more of the emission into the wings. We have supposed that the star+wind complex is seen edge-on, which is likely to be a crude approximation. However if the inclination angle would be changed, it would mainly change the filling factor. The latter has been written as a function r1(θ).r2(ϕ), r2(ϕ) being random and r1(θ) being composed with cosine functions. With a different inclination angle, the maximum of r1(θ) would not be reached at the equator, as in our simple model.

Lamers: Is there any evidence for corotating interactive regions in B[e] stars?

Bouret: I do not know anything about the CIR's in the B[e] stars. We are just interested in Herbig Ae/Be stars for the moment. However a model of the environment of B[e] star involving CIR's would be consistent if rotational modulation could be observed for some lines. I have only heard about models involving CIR's for O stars, Be stars and of course Herbig stars. So if CIR's are invoked in such different stars, there is no reason to rule them out for B[e] stars.

Owocki: What are the wind flow speeds behind the CIR shocks. If they are only 100 km.s^{-1} then this implies shocks that are probably too weak to produce X-rays observed by ROSAT.

Bouret: We have assumed a velocity difference between the fast and the slow streams of 100 km.s^{-1}, as inferred from the rotational modulation observed for the Mg II absorption lines for example. With a mass loss of about 1.2 10^{-8} $M_\odot$ yr^{-1} usually considered for AB Aur, it is indeed impossible to explain the X-ray luminosity measured by ROSAT (L_x=3.3 10^{23} erg.s^{-1} in the 0.1-2.4 keV range). However if the shocks at the boundaries of CIR's are MHD shocks (it is the case for the sun), mechanisms like the dissipation of Alfvén waves or ohmic heating must take place, and a magnetic field intensity of about 100 G, which corresponds to the equipartition field at the stellar surface constitute then a confortable reservoir of energy.

Zickgraf: The energy range of ROSAT is 0.1 to 2.9 keV. May be 0.1 keV

is soft enough?
Owocki: Maybe, but just barely if shocks have $\Delta V \leq 100$ km.s^{-1}
Pogodin: What can you say about the variability of the N V line?
Bouret: No information exists now.

NEW SPECTROSCOPIC RESULTS FOR THE GALACTIC B[e]-SUPERGIANT MWC 314

Y. FRÉMAT
Université de Mons-Hainaut, Groupe d'Astrophysique et de Spectroscopie, 20 place du Parc, 7000 Mons, Belgium

A.S. MIROSHNICHENKO
Central Astronomical Observatory of the Russian Academy of ciences at Pulkovo, Saint-Petersburg, 196140 Russia

AND

L. HOUZIAUX
Université de Mons-Hainaut, Groupe d'Astrophysique et de Spectroscopie, 20 place du Parc, 7000 Mons, Belgium

1. Introduction

$MWC314 = BD{+}14°3887$ was discovered by Merrill (1927) who found hydrogen and Fe II emissions in its spectrum. Very few spectroscopic observations of this object are available in the literature. Recently, Miroshnichenko (1996) reported results of its new photometric and medium–resolution (~ 2 Å) spectroscopic study. Analysing the SED and Balmer emission line profiles he concluded that MWC 314 is a highly reddened supergiant with a strong and slow stellar wind which could be considered as an LBV candidate.

2. Observations

New spectroscopic observations were obtained at the 1.52 meter telescope of the Observatoire de Haute-Provence on June 12, 1996 using the AURELIE spectrograph in two spectral regions, 4280 - 4720 and 6150 - 7030 Å. Our spectral resolution was 0.6 Å in the H_γ region and 1.1 Å in the H_α region. A lot of interstellar absorption features were found but no photospheric lines appeared and no P Cygni type absorptions were observed.

A. M. Hubert and C. Jaschek (eds.), B[e] Stars, 97–99.

3. Balmer lines

A possible solution to explain the absence of P Cygni type absorptions is an axisymmetric model in which the circumstellar gaseous disk formed by the stellar wind is seen at a certain inclination angle i (e.g. Pogodin 1992) relatively to the line of sight which is greater than the disk viewing angle γ. In this way, the best fit to the observed Balmer line profiles and infrared excess is reached with the following parameters : T_*=25000 K, $T_{wind} = 0.8T_*$, $R_* = 50R\odot$, $\dot{M}$ =4 $10^{-5}M\odot\ yr^{-1}$, v_∞=500 km s^{-1}, $i = 55°$ and $\gamma = 40°$. Our calculations based on a Sobolev NLTE code (see Pogodin 1989) fit well the profiles observed in 1991 at the 6–meter Russian telescope and those of 1996 obtained with a higher spectral resolution.

4. The λ6678 He I line

We tried to check our results by fitting the profile of the λ6678 He I line which forms near the base of the wind. The neutral helium line profiles were computed using the comoving frame method (Mihalas & Kunasz 1978) in a spherically symmetric envelope using the above stellar and wind parameters, except $\dot{M}$ and T_* which we varied. The best fit was obtained with T_* = 25000 K and $\dot{M}$ =9 $10^{-6}M\odot\ yr^{-1}$. In his study on MWC 314 hydrogen Balmer lines, Miroshnichenko (1996) derived T_* = 30000 K assuming a spherically symmetric envelope.

5. Conclusions

New modeling of the MWC 314 Balmer line profiles put constrains on geometry of its circumstellar envelope. Our calculations of the λ6678 He I line profile shows that a region of the envelope where it forms could be assumed to be spherically symmetric. The difference in derived values of $\dot{M}$ could be due to the model simplifications. There was no evidence of the presence of the λ 4686 He II line in our spectra. This line should appear only when effective temperature reaches 27000 K (Schmutz et al., 1991). The stellar wind parameters derived from the Balmer line and the λ6678 He I line modeling agree with this limit.

References

Merrill, P.W. (1927), *ApJ*, **65**, 286
Mihalas, D. & Kunasz, P.B. (1978) *ApJ*, **219**, 635
Miroshnichenko, A.S. (1996) *A&A*, **312**, 941
Pogodin, M.A. (1989) *Afz*, **31**, 150
Pogodin, M.A. (1992) *Sov.Astr.Let.*, **18**, 1066
Schmutz, W., Leitherer, C., Hubeny, I., Vogel, M., Hamann, W.R., & Wessolowski, U. (1991) *ApJ*, **372**, 664

Discussion

Cassinelli: You say that the lack of He II 4686 means that T < 27 000 K. I would have thought that one would need T $\geq$ 50 000 K to form He II 4686. Could you comment further on the He II 4686 constraint?

Frémat: Schmutz et al (1991) studied the LMC B[e] supergiant R 84 and modeled H I, He I and He II lines in its spectrum. They found that there is a strong dependence of He II line characteristics on T_*: they should appear in the spectrum of a star with T_* between 26 000 and 27 000 K. At lower temperatures they are not seen.

Stee: How is the He I 6678 line formed? Did you take into account recombination processes? Where is He I 6678 formed? I think that if He I 6678 is not formed by recombination processes in the He II region, the temperature close to the star is too high to populate the corresponding level. Close to the star helium is once or twice ionized, not neutral.

Frémat: We took into account recombination processes. We think that the λ 6678 He I line is formed near the stellar surface at the beginning of the envelope. This could explain that the λ 6678 emission line observed in MWC 54 is not symmetrical. Further, a computation shows that the profile is more sensitive to a variation of photospheric velocity than to a variation of terminal velocity.

HIGH-RESOLUTION SPECTROSCOPY OF HD 45677

G.L. ISRAELIAN

Insituto de Astrofisica de Canarias, E-38200 La Laguna, Tenerife, Spain

AND

F. MUSAEV

Special Astrophysical Observatory of the Russian AS, Nizhnij Arkhyz 357147, Russia

1. Introduction

We present high-resolution spectra of the peculiar B[e] star HD 45677 (FS CMa) obtained at 3 observing nights. In our paper we focus on the variability of Balmer and neutral helium lines.

A detailed description of HD 45677 has recently been provided by Sitko et al. (1994), Grady et al. (1993), de Winter & van den Ancker (1997) and Israelian et al. (1996) and we presently refrain from it. HD 45677 suffers from an ongoing large deep minimum, which started around 1950 with a detected range $V = 7^m.2 - 8^m.8$. The minimum has been explained (de Winter & van den Ancker 1997) by an obscuration effect probably due to large ($> 1\mu$m) circumstellar dust grains which were created by an evaporization of a large cometary body or by blown out material around 1950. The detected accretion can be the origin of a bipolar flow (Schulte-Ladbeck et al. 1993), which could be the region containing small grains. Spectroscopic properties of HD 45677 have been discussed in detail by Swings (1973) who originaly suggested that HD 45677 is a young post-main-sequence object surrounded by a dust shell, left over from formation of the star. Double-peaked Fe II emission lines indicate the presence of a gaseous disk around the star.

A. M. Hubert and C. Jaschek (eds.), B[e] Stars, 101–103.

2. Discussion

Our recent observations of Hβ clearly show 3 absorption components. The first one at zero velocity has been reported already by us and other observers and it is most probably formed in the edge-on disk. The second variable component is located at about $-40\ km\ s^{-1}$ and has never been mentioned in the literature. The third absorption component appears broad and blue-shifted and is located at $-140\ km\ s^{-1}$. We clearly record a large variability within 5 days. It appears as if some absorption remains at $-140\ km\ s^{-1}$ and an extra emission showed up in the blue together with a new absorption component at about $-100\ km\ s^{-1}$. So we do not observe any motion of the component located at $-140\ km\ s^{-1}$.

These new observations confirm our previous estimate (Israelian et al. 1996) of the rotational velocity of $v \sin i = 70\ km\ s^{-1}$. We succeeded to fit the three helium lines at 4471, 4387 and 5876 Å because the veiling in these spectra was not as strong as for January 11-14 1995 (Israelian et al. 1996). We again observe day to day variations of the helium, Balmer, Mg II lines. Because of the co-variations of the emission/absorption components of the Balmer, Mg II and He I lines, one could suggest that they have the same origin. If we want to link the variations of helium and Mg II lines with those seen in Hβ, we have to assume immediately that the gaseous cloud (or at least some part of it) has accreted and gave rise to the inverse P Cygni profile of the Mg II line and dumped more neutral helium in the atmosphere by which the veiling has diminished. Alternatively, we can think of some other (yet unknown) mechanisms responsible for the variations of the helium and Mg II lines. Cool gas is clearly observed in the line-of-sight. The outer envelope is clumpy and the motion of the gaseous clouds gives rise to the sharp absorption components visible in the Na, S and O emission lines. Details of these observations and their possible interpretation were a subject of the study by Israelian & Musaev (1997).

References

Grady, C.A., Bjorkman, K.J., Shepherd, D., Schulte-Ladbeck, R.E., Pérez, M.R., de Winter, D., Thé, P.S. 1993, ApJ 415, L39

Israelian, G., Friedjung, M., Graham, J., Muratorio, G., Rossi, C., de Winter, D. 1996, A&A, 311, 643 (Paper I)

Israelian, G., Musaev, F. 1997, A&A, in press

Schulte-Ladbeck, R.E., Shepherd, D.S., Nordsieck, K.H., Code, A.D., Anderson, C.M., Babler, B.L., Bjorkman, K.S., Clayton, G.C., Magalhães, A.M., Meade, M.R., Taylor, M., Whitney, B.A. 1993, ApJ 401, L105

Sitko, M.L., Halbedel, E.M., Lawrence, G.F., Allyn Smith, J., Yanow, K. 1994, ApJ 432, 753

Swings, J.P. 1973, A&A 26, 443

de Winter, D., van den Ancker, M.E. 1997, A&AS, 121, 275

Discussion

Bjorkman: Is there any correlation between the appearance of inverse P Cygni type profiles in the UV lines (Mg II) and the appearance of red-shifted absorption in the He line peaks?
Israelian: All we observe is that Balmer lines vary simultaneously with He I and Mg II (4481 Å). We were however unable to link these two types of variability. They have a very complex character. Obviously we need to monitor this object in order to understand the nature of the spectral variability.
Owocki: It is interesting to note that dynamical WCD models show material flowing both outward from the outer disk and inward from the inner disk. Your observations are time-variable and so cannot be completely explained by such a steady WCD model, but it might be interesting to see if this overall flow configuration could reproduce your inverse P Cyg type profiles.
Israelian: Yes, it would be interesting to check wether WCD models can be applied to HD 45677 in order to explain, and if possible, to fit the inverse P Cyg profile of Mg II 4481 and the P Cyg profile of H_{β}.

THE GALACTIC B[e] SUPERGIANT CANDIDATE HD 87643

RENÉ D. OUDMAIJER
Imperial College of Science, Technology and Medicine,
Blackett Laboratory, Prince Consort Road,
London, SW7 2BZ, U.K.

1. Introduction

HD 87643 has long been known as a peculiar B-type emission line star. Its spectrum is dominated by emission lines from FeII and the Balmer series. Of particular note is the P Cygni shape of the hydrogen and helium lines. Except for the blue-shifted absorption components and interstellar absorption lines, no absorption lines have been found.

From our own high resolution spectra around the Na D lines, we find that the velocities of the narrow interstellar absorption lines agree with the line of sight velocities towards the Carina arm. HD 87643 is located in the tangent point of the Carina arm, and the kinematic evidence suggests it is located beyond the local material, placing it beyond 1 kpc. This implies a luminosity comparable to those of B supergiants, rather than those of (pre-)Main Sequence stars. The rough distance estimate, combined with its spectral characteristics, suggests that HD 87643 is a Galactic counterpart of the B[e] supergiants in the Magellanic Clouds.

2. Spectropolarimetry

We have started a spectropolarimetric observing programme of B-type emission line stars, and report on our January 1997 Anglo-Australian Telescope observations of HD 87643. The observed spectrum is shown in the left hand plot of Figure 1. The polarization is enhanced in the P Cygni absorption and along the emission profile. Since the polarization changes across Hα, one can already conclude two things; firstly, the main mechanism that causes the polarization changes is electron-scattering, rather than scattering by small dust particles, and secondly that the electron-scattering region deviates from circular symmetry on the sky.

A. M. Hubert and C. Jaschek (eds.), B[e] Stars, 105–107.

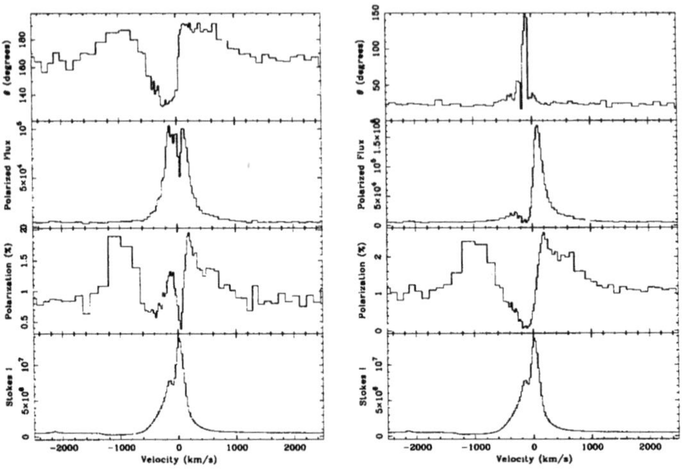

Figure 1. The polarization spectrum around Hα. The left hand side shows the observed spectrum, and the right hand side the spectrum corrected for interstellar polarization. The data has been rebinned such that individual pixels have 1σ uncertainties in polarization of 0.1%

Correction for the interstellar and circumstellar dust- polarization is not trivial. After considering many different methods, we present a preliminary result of the most promising; assuming that the Hα line is un-polarized at the line-centre, one can use the observed polarization as the value for the inter- and circumstellar polarization. It is hard to measure a value from the polarization spectrum (Fig. 1), due to the strong changes across the line. However, in QU-space it is easier to find, and we have adopted a value (Q=0%, U=−1.2%) for the line-center polarization. The resulting intrinsic polarization spectrum is shown in the right hand figure. The behaviour is fairly unique, although the line is (by implication) un-polarized at the line center, the emission line wings are enhanced in polarization. Observationally only few instances of enhanced line-polarization are known, but not as pronounced as here. The literature on such behaviour is very terse, Wood, Brown & Fox (1993, A&A 271, 492) calculated the polarization of circumstellar disks. Their model runs with rotating and expanding disks can provide polarization spectra very similar to what is observed in HD 87643. Of course, more modelling needs to be done, in order to quantify the properties of the electron-scattering envelope of HD 87643, yet the agreement between the observations and the published schematic models by Wood et al. is very encouraging. More details of our study of HD 87643 can be found in Oudmaijer, Proga Drew & de Winter (1998, MNRAS submitted).

Discussion

Friedjung: If I understand you have a non sperically symmetric axisymmetric wind. Is it in something like a disk? What is the opening angle of the wind region?
Oudmaijer: The opening angle of the wind is given by the viewing angle subtended by the star's equator and one of the most extreme points of the envelope. The absorption feature of the P Cyg profile disappears when the inclination angle becomes greater than the viewing angle.

RAPID SPECTRAL VARIABILITY AND SPATIAL AND KINEMATICAL STRUCTURE OF CIRCUMSTELLAR ENVELOPES OF SOME Ae/Be STARS

M.A. POGODIN
Pulkovo Observatory, 196140 Pulkovo, St.Petersburg, Russia

This investigation was performed within the frame of a complex program on the study of structural and kinematical peculiarities in the gaseous circumstellar envelopes around peculiar Ae/Be stars. The method involved is based on a search for rapid line-profile variability of lines originating in different circumstellar regions. Analysis of these variations opens the way for reconstructing the density and velocity distributions throughout the envelope.

More than three hundred and fifty high-resolution CCD spectra for four peculiar Ae/Be stars have been collected at the ESO (Chile, CAT-1.4m) and CrAO (Crimea, 2.6m Shajn telescope) in 1994–1997. The following objects were chosen for investigation. HD 50138 and HD 31648 are members of the peculiar Bep stars catalog by Allen & Swings (1976). They exhibit some features typical for both evolved Bep stars and early-type objects (Thé et al. 1994, Grady et al. 1994). One of the poorly-studied emission line objects HD 36112 was included to the new catalog of young Herbig Ae/Be stars (Herbig 1960) by Thé et al. (1994). HD 100546 was proposed by Hu et al. (1989) as a candidate Herbig star. Analysis of observational data allowed to reveal some properties which are likely to be common for the circumstellar media near these stars.

The Balmer line profile variations within one night in all the objects under investigation are observed in the form of "standing waves" on the residuals. They are likely to be a common type of the rapid variability in the Herbig Ae/Be stars. Beskrovnaya et al. (1995) proposed an interpretation of this phenomena on the basis of numerical modeling of an envelope containing a jet-like inhomogeneity (Pogodin 1990). It has been found that the rotation of a long-lived jet around the star results in the local intensity changes in the line profile without noticeable positional shift on the

A. M. Hubert and C. Jaschek (eds.), B[e] Stars, 109–110.

timescale of a few hours. This is connected with the fact that near the line-of-sight the jet is moving along the surfaces of equal radial velocities.

The behaviour of He I 5876 line profile in the spectra of HD 100546, HD 31648 and HD 50138 cannot be explained in terms of pure wind or pure infall. In the circumstellar gas of these objects the signs of both the stable outflow (weakly variable blue peak) and non-stationary accretion (highly variable red part of the profile) have been observed. Hence, there exists the kinematical stratification of the circumstellar gas around Herbig Ae/Be stars (Pogodin 1997; Vieira et al. 1997).

Violent asymmetry of the DNa I line profile observed in HD 36112 and HD 50138 with the blue part being more developed can be connected with existence of optically thick disk in the remote part of the envelope. Such disk may be either of protostellar origin or a result of the presence of low-velocity wind (Beskrovnaya et al. 1997b). It seems probable that there exists a region of the wind with low initial velocity which cannot leave the system and is likely to support the outer shell. Interaction of such a shell with faster wind can result in spatial inhomogeneity of the envelope, as well as in the loss of angular momentum by some portion of the circumstellar gas rotating around the star, leading to its subsequent infall onto the stellar surface. This hypothesis can explain the spectral peculiarities observed in the objects of investigation but the problem of their evolutionary status remains open.

Acknowledgements

This work was sponsored by the INTAS under grant 93-2478-ext.

References

Allen D.A. and Swings J.P. (1976), *A&A*, **vol. no. 47**, p. 293
Beskrovnaya N.G., Pogodin M.A., Najdenov I.D., Romanyuk I.I. (1995), *A&A*, **vol. no 298**, p. 585
Beskrovnaya N.G., Pogodin M.A., Yudin R.V., Franco G., Vieira S., Evans A. (1997), *A&AS*, in press
Beskrovnaya N.G., Pogodin M.A., Miroshnichenko A.S. et al. (1997), *A&A*, to be submitted
Grady C.A., Perez M.R., The P.S. (1994), *ASP Conf. Series*, **vol. no 62**, p. 409
Herbig G.H. (1960), *ApJS*, **vol. no 4**, p. 337
Pogodin M.A. (1990), *Astrofizika*, **vol. no 32**, p. 371
Pogodin M.A. (1997), *A&A*, **vol. no 371**, p. 185
Thé P.S., de Winter D., Perez M.R. (1994), *A&AS*, **vol. no 104**, p. 315
Vieira S., Pogodin M., Franco G. (1997), *A&A*, in preparation

INFRARED PROPERTIES OF B[e] STARS

L.B.F.M. WATERS
Astronomical Institute, University of Amsterdam
Kruislaan 403
1098 SJ Amsterdam, The Netherlands

SRON Laboratory for Space Research Groningen
P.O. Box 800
9700 AV Groningen, The Netherlands

AND

P.W. MORRIS, R.H.M. VOORS AND H.J.G.L.M. LAMERS
SRON Laboratory for Space Research Utrecht
Sorbonnelaan 2
3584 CA Utrecht, The Netherlands

Abstract. We discuss the infrared properties of B[e] stars and related objects, with emphasis on recent results obtained with the Short Wavelength Spectrometer (SWS) of the Infrared Space Observatory (ISO). The B[e] stars show evidence for the presence of a hot, and in many cases also a cool dust continuum with sometimes prominent solid state features. In most cases the dust is oxygen-rich, but carbon-rich components can also be present. The two B[e] supergiants GG Car and HD 87643 show a large difference in properties at IR wavelengths, perhaps suggesting that different evolutionary scenarios can lead to the B[e] supergiant classification. GG Car *may* have C-rich dust. The IR properties of B[e] stars are compared to those of three other groups of hot stars with circumstellar matter: 'classical' Be stars, Luminous Blue Variables, and Herbig Ae/Be stars.

1. Introduction

The classification of B type stars with emission from both Hα and forbidden lines in their optical spectra as B[e] stars has resulted in a very heterogeneous sample of hot stars with circumstellar matter (CSM). Since

A. M. Hubert and C. Jaschek (eds.), B[e] Stars, 111–126.

hot stars can have a substantial amount of CSM in very different evolutionary phases, such a spectroscopic and rather phenomenological classification is likely to contain stars in evolutionary phases ranging from pre-main-sequence to post-main-sequence, and with a wide range in luminosities (and hence mass).

Therefore it is important to find additional criteria for the proper classification of B[e] stars and to pinpoint their evolutionary status. During this workshop, it has become evident that within the class of B[e] stars several distinct groups can be identified, with more or less clear evolutionary status. In the contribution of Lamers et al. (these proceedings), an overview of the different groups of stars within the umbrella of B[e] stars has been given, with their characteristics; we mention the B[e] supergiants, massive post-main-sequence stars with strong, axisymmetric winds, and possibily related to the Luminous Blue Variables (e.g. Zickgraf et al. 1986; see also Zickgraf, these proceedings), the B[e] stars in the class of pre-main-sequence Herbig Ae/Be stars, the class of hot post-AGB stars with detached AGB winds, the symbiotic stars, and finally a 'miscellaneous' group of stars with peculiar spectra.

The infrared spectral region is useful to determine the evolutionary status of objects with large amounts of CSM, since it is very rich in diagnostic tools, such as permitted and forbidden emission lines, many molecular bands, and thermal emission from circumstellar dust. Indeed, one of the main characteristics of B[e] stars is their strong emission at IR wavelengths due to the presence of large amounts of circumstellar dust (Allen 1973; Allen & Swings 1976). While the mere presence of circumstellar dust cannot easily be used to distinguish different sub-groups within the class of B[e] stars, the shape of the energy distribution, *IR spectroscopy of solid state features*, and direct imaging of B[e] stars at (far-) infrared wavelengths can be used to determine the compositon and geometry of the circumstellar dust. This in turn can constrain the evolutionary status of B[e] stars.

This review deals with the IR properties of B[e] stars, in particular with the new results obtained with the Infrared Space Observatory (ISO; Kessler et al. 1996) and its Short Wavelength Spectrograph (de Graauw et al. 1996). In Sect. 2, we discuss the IR properties of some B[e] supergiants, in Sect. 3 we briefly describe the ISO-SWS spectra of objects that are related to the B[e] supergiants (sometimes also classified as B[e] stars): the 'classical' Be stars, the Luminous Blue Variables, and the Herbig Ae/Be stars. Sect. 4 summarizes this review.

2. Infrared properties of B[e] supergiants

As mentioned in the introduction, a class of B supergiants with very strong Hα emission (equivalent widths often exceeding 200 Å) and forbidden line emission from low-ionization species such as Fe II and O I is part of the group of B[e] stars. These stars were first recognized as a group by Zickgraf et al. (1986), who studied a sample of hot emission line stars with dust shells in the Magellanic Clouds. Stars in this group show evidence for a rotating and slowly expanding disc in the low-excitation lines, and for a fast, radiation-driven wind in high-excitation ultraviolet resonance lines. The model proposed by Zickgraf et al. (1986) for the wind geometry of these B[e] supergiants resembles that of the classical Be stars (which are dwarfs or giants).

The B[e] supergiants are believed to be rapid rotators, with a two-component wind: an equatorial region in which the density is high and outflow velocities are low, and a polar region in which the densities are low and the outflow velocities high. This wind geometry may be the result of the bi-stability of hot star winds (Lamers & Pauldrach 1991; Lamers et al. 1995), in which the wind acceleration is a sensitive function of the ionisation of the wind. This ionisation in turn depends on the temperature of the star, and, through the von Zeipel effect, has a latitude dependence for rapidly rotating stars.

The temperatures and densities in the equatorial part of the wind can reach values that allow the formation of molecules (e.g. CO) and eventually dust condenses in the outflow. Indeed, CO has been found in the spectra of several B[e] supergiants (McGregor et al. 1988; Morris et al. 1996; for a review, see Morris et al. 1997), and they show ample evidence for the presence of hot ($\approx$ 1000 K) as well as cool ($\approx$ 100 K) dust (Allen 1973; Zickgraf et al. 1986; McGregor et al. 1988). Spectroscopy of the dust is a useful tool to determine the chemical composition of the material in the disc, e.g. whether the star has an O-dominated or C-dominated chemistry.

In Fig. 1 we show the ISO-SWS spectra of GG Car and HD 87643, two 'representative' members of the group of B[e] supergiants (note that Lamers et al. (these proceedings) classify HD 87643 as a B[e] supergiant, but with some doubt). It is immediately apparent from Fig. 1 that B[e] supergiants show a wide range in behaviour at IR wavelengths, both in the shape of the energy distribution as well as in the appearance of solid state features. We discuss these two cases separately below.

GG Car: this object is a single-lined eclipsing binary with an orbital period of 31 days and an eccentricity of about 0.3 (Gosset et al. 1985 and references therein). The optical spectrum suggests a spectral type B0-B2 with emission from (among others) Balmer lines, FeII and [FeII] (e.g. Lopes et al. 1992).

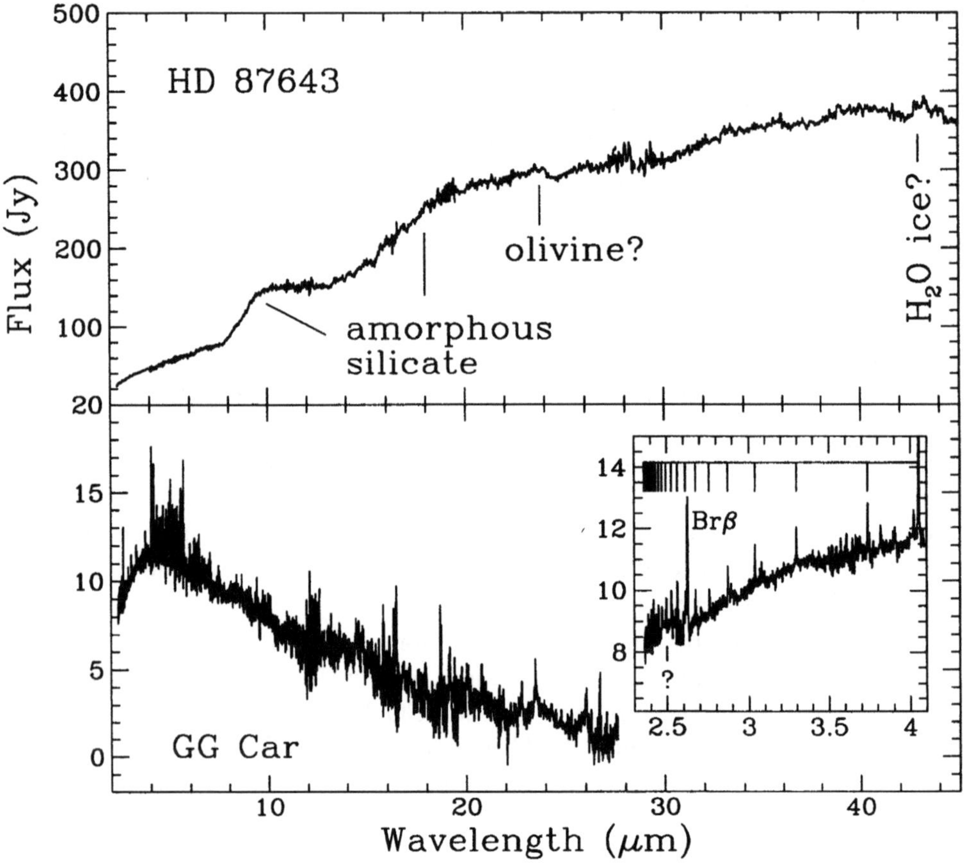

Figure 1. 2.4-45 μm ISO-SWS spectra of two B[e] supergiants, GG Car and HD 87643. Notice the strong dust continuum in both stars. In HD 87643 crystalline silicates may be present, and there is evidence for crystalline H_2O ice. GG Car shows a featureless hot dust continuum. Notice also the strong emission from the Pfund and Brackett lines of hydrogen

The Na I doublet has a blue-shifted absorption component at -125 km/s, indicating that the low-excitation envelope is expanding. The Balmer lines show P Cygni profiles. There is no evidence for a nebula (Thackeray 1950).

The SWS spectrum shows a hot dust continuum with very little structure. The 9.7 μm silicate feature is not obvious in our spectrum. Ground-based 10 μm narrow-band photometry presented by McGregor et al. (1988) suggests that a weak 10 μm feature may be present, which we would have detected. There is no evidence for cool dust, nor for emission from the C-rich UIR bands at 3.3, 6.2, 7.7, 8.6 and 11.3 μm. The shape of the energy distribution is compatible with dust formation in a wind which flows at constant velocity (r^{-2} density gradient). Modelling of the SWS spectrum indicates that dust shells consisting mainly of amorphous silicates do not give a good fit, since this would result in too strong emission in the 9.7 μm Si-O stretch band. This may imply that for some unknown reason the grains that condense in the outflow of GG Car are much larger than those typically found in dusty winds, or that the dust is C-rich. Indeed, models with *amorphous carbon* do much better. The presence of C-rich dust is difficult to understand, since a C-rich chemistry (and hence C-rich photosphere) would be required. The abundant presence of hydrogen recombination lines (Fig. 1) suggests that the star has not yet evolved to a C-rich and H-poor Wolf-Rayet star. This issue remains to be resolved. The SWS spectrum shows a bump at about 2.55 μm with a width of 0.1 μm, which is probably instrumental.

Morris et al. (1996) demonstrate from near-IR groundbased spectra the presence of CO first overtone ro-vibrational emission, and the SWS spectra confirm its presence. The CO is highly excited and presumably located close to the star. The CO fundamental ro-vibrational band at 4.66 μm is not seen, but this may be due to veiling by the hot dust continuum. No other molecules are detected in GG Car. The spectrum shows emission lines from hydrogen (Brβ, Brα and many lines from the Pfund series). The line strength ratios suggest that they are optically thick, which is similar to the hydrogen recombination spectrum of 'classical' Be stars (see below). It is very well possible that, as in the case of the Be stars, the bulk of the line emission in the hydrogen recombination lines of GG Car orginates from a dense, optically thick disk.

Several other B[e] stars have IR spectra similar to GG Car, such as CPD-57° 2874 and HD 327083.

HD 87643: The optical spectrum (spectral type B3; Shore et al. 1990) is dominated by numerous emission lines and prominent P Cygni profiles in the Balmer lines. These Balmer lines also have double-peaked emission, perhaps indicating a rotating component. Hβ has an edge velocity of -1340 km/s (Lopes et al. 1992). Long-term variability in emission line strength and/or continuum is observed, but no periodicity has been found so far (de Freitas Pacheco et al. 1985; Lopes et al. 1992). Mass loss rates quoted in the literature show a large range; Lopes et al. (1992) derive $\dot{M} \approx 10^{-5}$ $M_\odot$/yr, based on Hβ. If the mass loss rate is as high as this, it suggests a high luminosity for HD 87643. The distance to HD 87643 is uncertain; Lopes et al. (1992) derive 2.9 kpc based on the Na I interstellar absorption line strength. The reflection nebula near HD 87643, first described by Henize (1962), is irregular in shape (Surdey et al. 1981) and has dimensions of the order of 1 arcmin.

The continuum energy distribution between 2.4 and 40 μm continuously rises, and levels off at $\lambda \approx 45$ μm, indicating the presence of a wide range of dust temperatures. This shape is incompatible with a simple outflowing wind at constant velocity. It can be explained by a wind in which the mass loss rate was much higher in the recent past, or by a dusty disc.

The SWS spectrum shows a rich structure of solid state features, in particular prominent emission from circumstellar amorphous silicates at 9.7 and 18 μm (Fig. 1). The shape of the 9.7 μm feature is somewhat peculiar, and may be affected by absorption from interstellar amorphous silicates. Some weaker emission bands are detected at longer wavelengths, that may be identified with crystalline silicates (pyroxenes and/or olivines). Crystalline silicates are ubiquitous in cool circumstellar dust envelopes of stars in very different evolutionary phases, such as pre-main-sequence Herbig Ae/Be stars (Waelkens et al. 1996), Luminous Blue Variables (Waters et al. 1997) and red supergiants and post-Asymptotic Giant Branch stars (Waters et al. 1996). Crystalline silicates therefore can occur in situations in which the source of circumstellar matter is outflowing winds as well as accretion. The spectrum shows a bump near 43-44 μm which is probably due to crystalline H_2O ice. The outer parts of the shell (disc?) may therefore contain grains covered with an icy mantle. There is no evidence for the presence of emission from the UIR bands.

2.1. A REMARKABLE B[e] SUPERGIANT IN WD 1

In a study of the stellar population of the heavily reddened cluster Wd1 at a distance of 5 kpc in the Ara constellation, Westerlund (1987) describes the spectrum of star 9 (star C in the nomenclature of Borgman et al. 1970) as an early B star with $\log(L/L_\odot) = 4.9$ and strong emission from Hα and forbidden lines. This description qualifies the star to be a member of the class of B[e] stars, and its location in the cluster HR diagram suggest it is a supergiant. Borgman et al. (1970) and Koornneef (1977) find a strong near-IR excess in the B[e] star, as well as emission from several M supergiants.

The B[e] star was recently studied in more detail at IR and radio wavelengths. We show in Fig. 2 the forbidden emission lines detected with ISO-SWS (Clark et al.; in preparation). Strong emission from [Ne III] and [S III] is seen as well as [O IV]. The latter line requires an excitation temperature of 80.000 K or more, which is clearly not compatible with the temperature of the B[e] star. We know of no other B[e] star with such high excitation for-

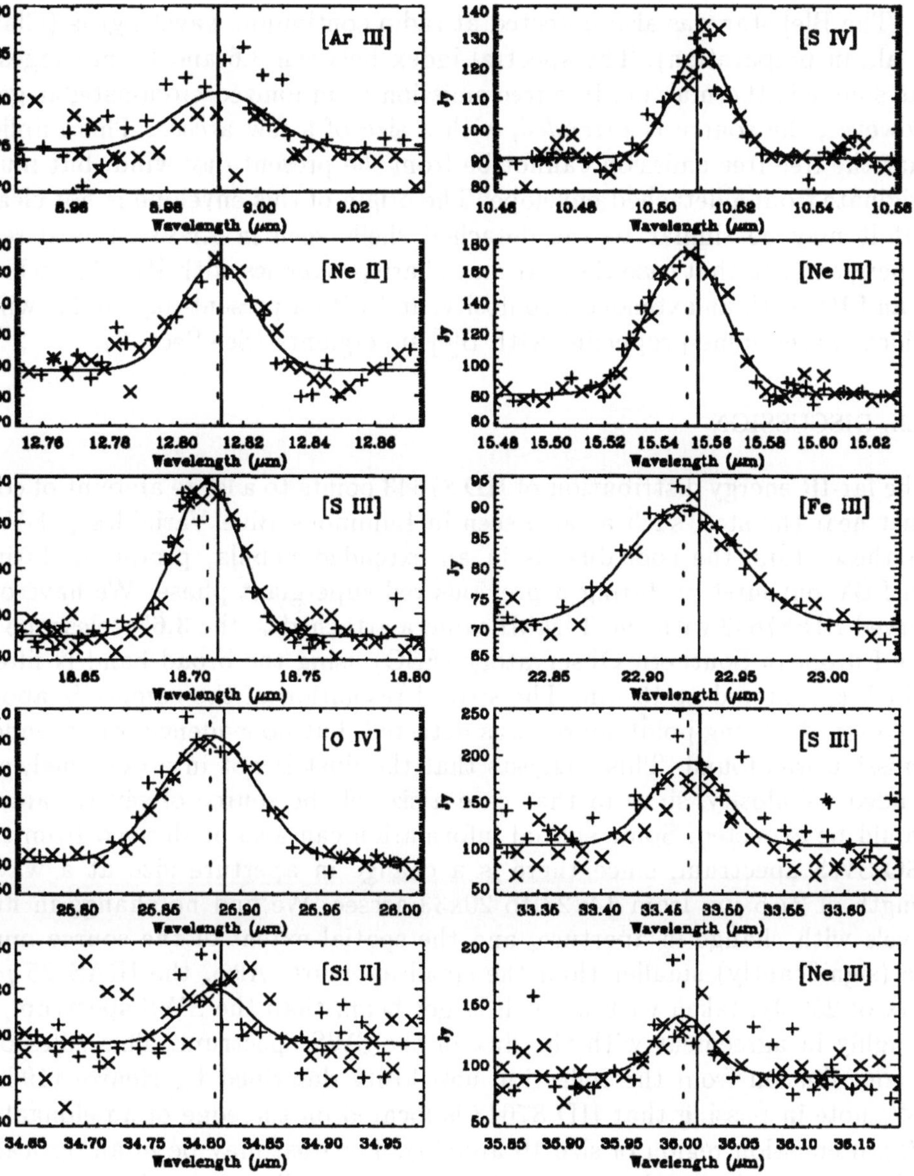

Figure 2. Emission lines observed in the ISO-SWS spectrum of the B[e] star in the young, heavily reddened cluster Wd 1 (Clark et al. in preparation). Notice the high excitation of the circumstellar gas, in particular the presence of [O IV] emission

bidden line emission. The origin of the high excitation lines may be shocked gas near the B[e] star, possibly resulting from a fast wind colliding with a slower one, or with the ambient interstellar medium.

The B[e] star was also detected at radio continuum wavelengths (Clark et al., in preparation). The spectral index between 3.6 and 6 cm suggests the source is thermal, i.e. free-free emission from ionized circumstellar gas. However, the source is *extended*, with a size of a few arcsec. This implies that the free-free emission cannot be from the present-day wind, but must originate from a detached envelope. The origin of this envelope is not clear, but it may be similar to the detached shells seen in LBVs or post-red-supergiants. In that case the star may share properties with HR Car, which is an LBV with an extended nebulosity, and with a present-day stellar wind which shares some properties with B[e] supergiants (See Sect. 3).

2.2. DISCUSSION

The far-IR energy distribution of HD 87643 points to a large amount of cool dust near the star, such as also seen in Luminous Blue Variables (LBVs). In these stars, the cool dust is in an extended nebula, produced during an LBV outburst or during a previous red supergiant phase. We have observed HD 87643 with the TIMMI camera attached to the 3.6^m telescope of the European Southern Observatory (ESO) using the broad-band N filter, which is centered at 10 μm. The spatial resolution of the camera is about 1 arcsec. A strong point source was detected, but no evidence for extended emission was found. This suggests that the dust is not in a detached, extended nebulosity, since in that case a size of the source of several arcsec would be expected. Some spatial information can also be derived from the ISO-SWS spectrum, since there is a change in aperture size at a wavelength of 27.5 μm from 14x27 to 20x33 arcsec. We find no change in flux levels with change in aperture, and the spatial extent of the source must be (significantly) smaller than the small aperture. Also, the IRAS 25 μm flux of 252 Jy, taken with a much larger beam than the SWS spectrum, is roughly in agreement with the flux in the SWS spectrum. This rules out a contribution from the reflection nebulosity described by Henize (1962). (We note in passing that HD 87643 is located on the edge of an elongated 60 μm emission region of size 10 arcmin). The dust may be in an optically thick, geometrically thin compact circumstellar disc. In such a geometry the temperature gradient of the dust is steeper than in spherical geometry, and the source size significantly smaller.

What is causing the large difference in IR properties of HD 87643 and GG Car? In the latter case, the energy distribution is compatible with dust condensing in an outflow (presumably in the equatorial regions of the wind) with an r^{-2} density gradient. Perhaps the mass loss rate of GG Car is smaller than that of HD 87643 (as suggested by the factor ≈ 2 difference in Hα equivalent width; note however that the Hα line strengths are variable), and the dust is optically thick in the case of HD 87643 while it is optically thin in GG Car. For an optically thick disc, the slope of the IR energy distribution at wavelengths shortward of the turndown is a function of the temperature gradient and not of the density gradient. There are several difficulties however with the interpretation of the IR spectrum of HD 87643 in terms of an optically thick disc. First, the required dust mass is very high, and difficult to estimate on the basis of the ISO-SWS data alone. Second, the 9.7 μm amorphous silicate feature is in emission, which in fact requires optically thin dust. However, there is evidence in the case of discs surrounding young Herbig Ae/Be stars that an optically thick disc can be present with the 9.7 μm silicate feature in emission (Mannings & Sargent 1997; see also Waters & Waelkens 1998).

It is interesting to point out the similarity between the IR properties of HD 87643 and those of HD 101584, a late B star in a wide binary system (orbital period 218 days) and strong optical emission lines from hydrogen and permitted and forbidden singly ionized Fe (Bakker et al. 1996a; 1996b). HD 101584 shows a strong P Cygni profile in the Hα line, indicating a dense stellar wind. The SWS spectrum of HD 101584 (not shown here) shows a steeply rising continuum, very similar to HD 87643, and a prominent peak due to crystalline H_2O ice. The model proposed by Bakker et al. (1996b) consists of an interacting binary with a prominent circum-binary disc and cooler material from a previous phase of mass loss and/or mass transfer. It is possible that a similar scenario also applies for HD 87643, in which the star is somewhat hotter than in HD 101584. Remarkably, GG Car, which is a confirmed binary, has an IR energy distribution compatible with outflow. If binarity affects the geometry of circumstellar matter, it is unclear why GG Car is so different from HD 87643.

3. Comparison with other hot emission line stars

3.1. CLASSICAL Be STARS

In general, classical Be stars are easily distinguishable from B[e] stars, because they do not show forbidden line emission, and because because Be stars have an excess at infrared wavelengths due to free-free and free-bound emission from ionized circumstellar gas, and lack circumstellar dust. This is evident from the shape of the IR energy distribution (e.g. Waters et al. 1987) and from IR spectroscopy. As an example, we show in figure 3 a small part of the ISO-SWS spectrum of the well-known Be star γ Cas (B0.5 IVe). The spectrum is dominated by numerous emission lines from high levels of

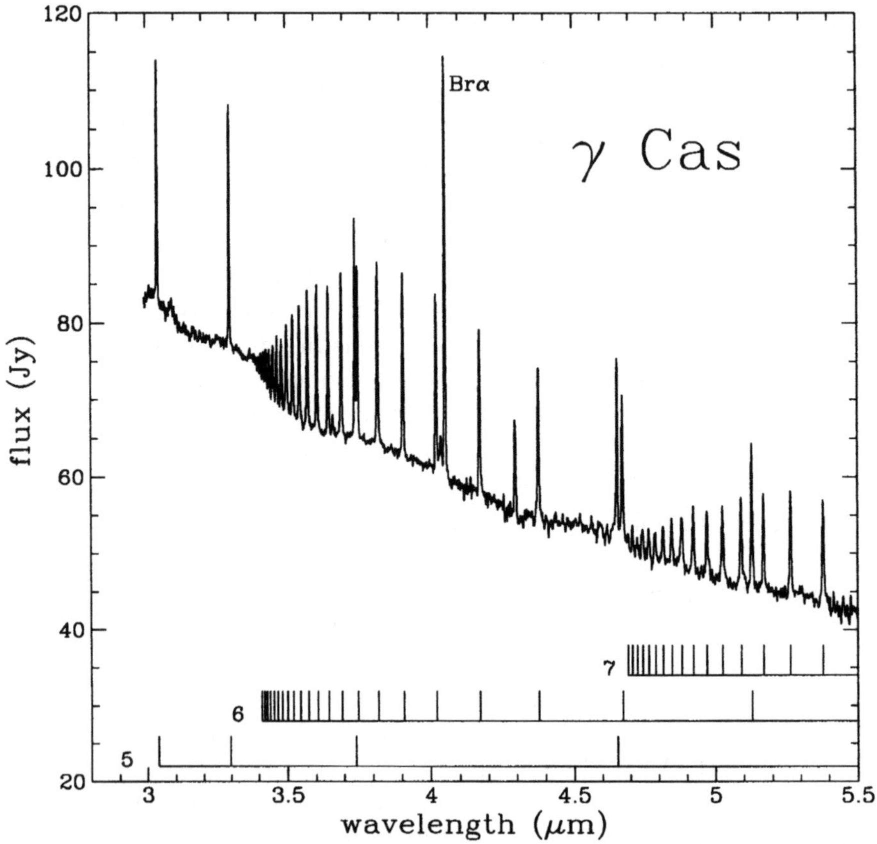

Figure 3. Part of the ISO-SWS spectrum of γ Cas, a 'classical' Be star. Notice the strong emission from HI recombination lines. Tickmarks indicate the positions of the lines of different series, which are connected by horizontal lines. The lower level of each series is indicated (5 = Pfund, 6 = Humphreys, 7 = Hansen-Brown)

hydrogen, typical for a recombination spectrum. We find no evidence for the presence of solid state dust features, such as the 9.7 μm silicate bump, or the UIR bands at 3.3, 6.2, 7.7, 8.6 and 11.3 μm. The conditions for the formation of circumstellar dust in Be star discs clearly are not favourable, perhaps due to the strong UV radiation field of the central star, which is not efficiently shielded by a large amount of circumstellar matter.

Nevertheless, the emission line strengths of the hydrogen recombination lines in γ Cas suggest that its circumstellar disc has very high densities, in excess of 10^{-11} g cm^{-3}. The line flux ratios suggest that an extended region with low densities is *absent.* This may be the reason why Be stars hardly ever show forbidden line emission in the optical, which requires extended low-density regions near the star with sufficient emission measure to produce detectable line emission.

3.2. LUMINOUS BLUE VARIABLES

The ISO-SWS spectra of LBVs show evidence for the presence of crystalline silicates (Lamers et al. 1996; Waters et al. 1997; 1998). The similarity between the solid state features seen in the dust shells surrounding LBVs and those observed in red supergiants led Waters et al. (1997) to suggest that some of these ring nebulae may be the result of a previous red supergiant wind and not of an LBV outburst. A similar suggestion was made by Smith et al. (1997) on the basis of abundance studies in LBV ring nebulae. This issue is not yet settled.

The discovery by Morris et al. (1996) of CO first overtone emission in the LBV HR Car, and its time variability (Morris et al. 1997) for the first time suggests a link between the class of LBVs and of B[e] supergiants. HR Car shows variability in spectral type (ranging between early-B and late-B) which is typical for LBVs. The presence of CO emission seems correlated with this variability, in the sense that CO emission is present when the star is hot. This can be explained qualitatively in terms of the wind bi-stability mechanism. When the star is hot, it has the smallest radius and therefore rotates fast. This allows the equatorial regions to develop a very slow wind in which molecules, and possibly dust can condense. When the star is cool, it has a more or less spherical wind in which the conditions are not favourable for the formation of molecules or dust.

HR Car has a compact, warm dusty detached dust shell with a complex geometry (Voors et al. 1997; see Fig. 4) as well as a more extended Hα nebulosity (e.g. Weis et al. 1997), pointing to multiple ejections of matter. The dust mass in the inner, compact nebula is less than 8 10^{-4} $M_\odot$ (Voors et al. 1997); both the geometry and the low amount of mass in the shell suggest that it is not the remnant of a red supergiant wind. The ISO-SWS spectrum (Fig. 4; Lamers et al. 1996; see also Waters et al. 1997) shows prominent amorphous silicates at 9.7 and 18 μm, presumably emitted by the compact, warm dust component. There is also evidence for crystalline silicates in the 30-45 μm region, emitted by cooler material which is probably associated with the more extended Hα nebula. The origin of this more extended nebula is not clear, it could be a previous LBV eruption or a remnant from a red supergiant phase. We note that HR Car lacks the emission band at 33.8 μm of Mg-rich and Fe-poor olivine, which has been seen in the LBV AG Car (Waters et al. 1997) and which is wide-spread in cool giants and supergiants with high mass loss rates (Waters et al. 1996). Instead, emission bands near 31, 32.5, 35, 38.8 (the latter marked with 'U'

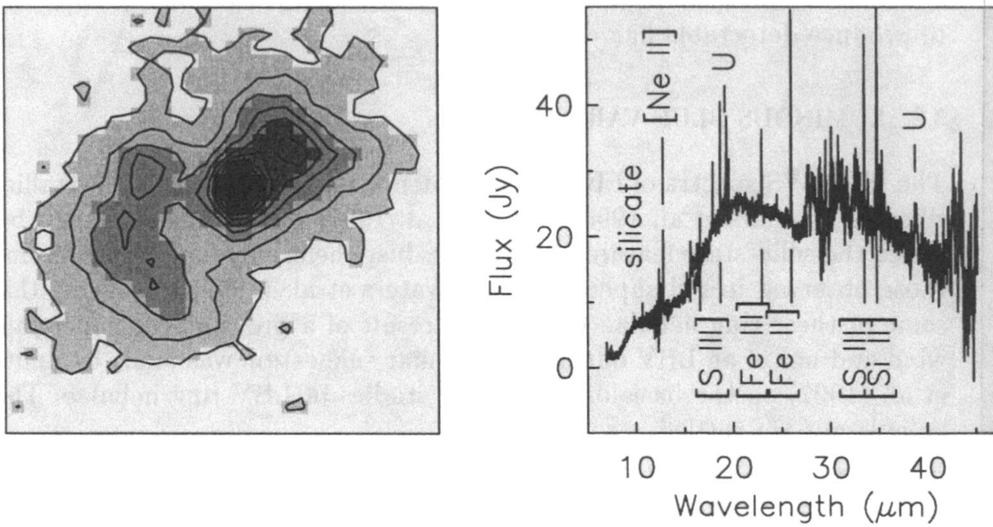

Figure 4. Left: ground-based 10 μm TIMMI (ESO 3.6m telescope) broad-band N image of the HR Car Nebula (Voors et al. 1997). The image covers 11x11 arcsec. The bright point source is the central star. Notice the non-axisymmetric distribution of dust. Right: ISO-SWS spectrum of HR Car, with the strongest forbidden lines indicated. Two narrow solid state features are unidentified and marked 'U'. The long-wavelength part of the spectrum shows evidence for crystalline pyroxenes but no strong emission from Mg-rich olivines is observed

in figure 4) and 42-44 μm are apparent (a second strong, narrow feature is seen at 19.8 μm, also labeled 'U'). The 35 μm peak could be due to Fe-rich crystalline olivine. Laboratory spectra of crystalline olivines show a shift to longer wavelengths with increasing Fe/Mg content (e.g. Koike et al. 1993). The other peaks are more difficult to identify, they poorly match laboratory spectra of crystalline pyroxenes available to us (Koike et al. 1993; Jäger et al., in preparation), although we do find weak emission at similar wavelengths in other oxygen-rich sources with crystalline silicates. It seems that HR Car has unusally high abundances of a particular type of crystalline silicate, and lacks others (such as Mg-rich olivines). This points to a different formation history of these grains.

3.3. HERBIG Ae/Be STARS

Herbig Ae/Be stars (hereafter HAEBE stars) are intermediate mass pre-main-sequence stars, and can be viewed as the higher mass counterparts to the T Tau stars (for recent reviews on this subject see Grady & Pérez (1998) and Waters & Waelkens (1998)). They are characterized by large amounts of circumstellar matter, in particular a strong excess due to warm

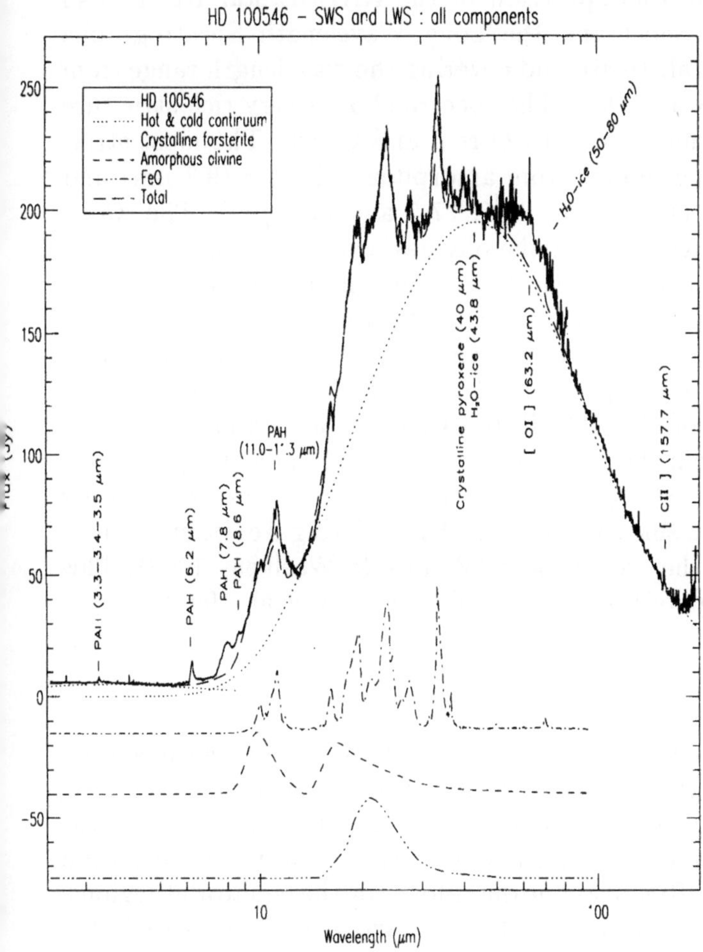

Figure 5. The ISO-SWS and LWS spectra of the HAEBE star HD 100546 (Malfait et al. 1998). Notice the rich structure in solid state features.

and cool circumstellar dust is usually observed (Hillenbrand et al. 1992 and references therein). The IR excess often starts at wavelengths between 1 and 2 μm, and the spectrum can either continue to rise to wavelengths beyond 50 μ, or shows a 'two-component' distribution with a dip around 10 μm. Recent imaging of Herbig Ae stars at millimeter wavelengths has demonstrated that the dust at least in some cases is in a flattened, disc-like geometry (Mannings & Sargent 1997; Mannings et al. 1997). These discs are possibly the site of on-going planet formation (Malfait et al. 1998). HAEBE stars are believed to be the predecessors of β Pic stars and of Vega type stars.

In Fig. 5 we show the ISO spectrum of the HAEBE star HD 100546 (B9Ve), taken with the Short and the Long Wavelength Spectrographs (SWS and LWS; Clegg et al. 1996), and covering the wavelength range from 2.4 to 200 μm (Malfait et al. 1998). The spectra show a very rich structure of solid state features, which are both O-rich and C-rich. The most striking features are the strong bumps from amorphous silicates (9.7 μm) and from crystalline olivines (16.5, 19.5, 23.5, 27.5 and 33.8 μm). The wavelengths of these bands suggests that the olivines are Mg-rich and Fe-poor, i.e. Forsterite. In addition, peaks near 43 and 60 μm are due to crystalline water ice, and a much weaker feature near 100 μm may also be due to this material. These spectral features are reminiscent of the composition of comets. Indeed, the ISO-SWS spectrum of comet Hale-Bopp shows a striking similarity with that of HD 100546 (Crovisier et al. 1997). At shorter wavelengths, emission from the UIR bands is observed, indicating the presence of C-rich dust. The ISO spectrum of HD 100546 is remarkable in its strength of the crystalline silicate features. Other Herbig Ae/Be stars show much weaker bands, or they are absent (Waters & Waelkens 1998). This may be related to the age of the star (van den Ancker et al. 1997).

4. Conclusions

The infrared properties of B[e] supergiants seem rather heterogeneous, if we use GG Car, HD 87643, HR Car and star 9 in Wd 1 as examples. Ignoring the last star for the moment, one may construct a picture in which the the B[e] supergiants are rapidly rotating massive stars with a flattened wind in which molecules, and often also hot dust, form in the equatorial regions, i.e. as proposed by Zickgraf et al. (1986). There are several moments in the evolution of massive stars when they can have such a geometry. (1) These stars may have evolved off the main sequence recently, and may have had a rapidly rotating O star as a progenitor (for example λ Ceph). In this case, no cool dust from a previous phase of high mass loss is expected. (2) They may have suffered extensive mass loss, either as an LBV or a red supergiant, and have lost a significant fraction of their H-rich envelope, exposing the (possibly rapidly rotating) interior layers. In this case a cool dust component may be present. An example is HR Car; perhaps star 9 of Wd 1 also falls into this category. At present, it is doubtful that HD 87643 fits in either group.

The other groups of stars discussed in this paper, Be stars and the young stars belonging to the group of Herbig Ae/Be stars, have infrared properies that are either very different from those of B[e] supergiants (the Be stars), or that are superficially similar (Herbig Ae/Be stars). In these young HAEBE stars, the dust is likely in a rotating disc which contains both oxygen-rich and some carbon-rich material, and a substantial cool dust component is present. The presence and spatial extent of this cool dust is perhaps a way to distinguish them from the B[e] supergiants with dusty equatorial winds.

Finally, it is well known that *binaries* in which one of the components is evolved can have a (stationary) rotating circumbinary disc containing sometimes a significant amount of mass. Examples of these categories are HD 101584 (late B; Bakker et al. 1996a, 1996b) and 3 Puppis (early A; Plets et al. 1995). These systems show IR spectra very similar to those seen in some B[e] stars, in particular HD 87643. The shape of the IR energy distribution of these systems depends on the radial extent of the (keplerian) disc, and is in general much broader than that of an 'outflow' dust disc with an r^{-2} density gradient. However, the case of GG Car illustrates that binarity does not always result in a broad IR energy distribution.

Acknowledgements. It is a pleasure to thank K. Malfait, C. Waelkens, M. van den Ancker, A. van Blokland, A. de Koter, Th. Lanz, J. Koornneef, S. Clark, R. Fender, and S. Dougherty for stimulating discussions on the B[e] stars discussed in this paper. LBFMW gratefully acknowledges financial support from an NWO *Pionier* grant.

References

Allen D.A., 1973, MNRAS 161, 145
Allen D.A., Swings J.P., 1976, A&A 47, 293
van den Ancker, M.E., Thé, P.S., Tjin A Djie, H.R.E., Catala, C., de Winter, D., Blondel, P.F.C., Waters, L.B.F.M.: 1997, A&A 324, L33-L36
Bakker E.J., Lamers H.J.G.L.M., Waters L.B.F.M., Waelkens C., Trams N.R., Van Winckel H.: 1996a, A&A 307, 869
Bakker E.J., Lamers H.J.G.L.M., Waters L.B.F.M., Waelkens C., 1996b, A&A 310, 861
Borgman J., Koornneef J., Slingenland J., 1970, A&A 4, 248
Clegg, P.E., et al., 1996 A&A 315, L38
Crovisier, J., Leech, K., Bockelée-Morvant, D., Brooke T.Y., Hanner, M.S. et al.: 1997, Science 275, 1904
De Freitas Pacheco J.A., Faria Lopes D., Landaberry S.C., Selvelli P.L., 1985 A&A 152, 101
Gosset, E., Hutsemekers, D., Surdej, J., Swings, J.-P.: 1985, A&A 153, 71
De Graauw, Th. et al.: 1996, A&A 315, L49
Grady, C., Pérez, M.R., 1998: Space Science Reviews, in press
Henize, K.G.: 1962, AJ 67, 612
Kessler, M.F., et al., 1996 A&A 315, L27
Koike, C., Shibai, H., Tuchiyama, A.: 1993, MNRAS 264, 654
Koornneef, J.: 1977, A&A 55, 469
Lamers, H.J.G.L.M., Pauldrach, A.W.A.: 1991, A&A 244, L5
Lamers, H.J.G.L.M. et al.: 1996, A&A 315, L225
Lamers, H.J.G.L.M., Snow, T.P., Lindholm, D.M.: 1995, ApJ 455, 269
Lopes, D.F., Damineli Neto, A., de Freitas Pacheco, J.A.: 1992, A&A 261, 482
Malfait, K., Waelkens, C., Waters, L.B.F.M., Vandenbussche, B., Huygen, R., de Graauw,

Th.: 1998, A&A (in press)
Mannings V., Koernier D.W., Sargent A.I., 1997, Nature 388, 555
Mannings V., Sargent A.I.,, 1997, ApJ 490, 792
McGregor, P.J., Hyland A.R., Hillier D.J., 1988, ApJ 324, 1071
Morris, P.W., Eenens, P.R.J., Hanson, M.M., Conti, P.S., Blum, R.D.: 1996, ApJ 470, 597
Morris, P.W., Voors, R.H.M., Lamers, H.J.G.L.M., Eenens, P.R.J.: 1997, in *Luminous Blue Variables: Massive Stars in Transition*, ASP Conf. Series vol. 120, Eds. A. Nota & H.J.G.L.M. Lamers, p. 20
Plets H., Waelkens C., Trams N.R., 1995, A&A 293, 363
Shore, S.N., Brown, D.N., Bopp. B.W., Robinson, C.R., Sanduleak, N., Feldman, P.D.: 1990, ApJS 73, 461
Smith L.J., Stroud M.P., Esteban C., Vilchez J.M., 1997, MNRAS 290, 265
Surdey, A., Surdey, J., Swings, J.-P., Wamsteker, W.: 1981, A&A 93, 285
Thackeray, A.D.: 1950, MNRAS 110, 524
Voors, R.H.M., Waters, L.B.F.M., Lamers, H.J.G.L.M., Trams, N.R., Käufl, H.U.: 1997, A&A 319, L21
Waters L.B.F.M., Cote J., Lamers H.J.G.L.M., 1987, A&A 185, 206
Waters, L.B.F.M., et al.: 1996, A&A 315, L361
Waters, L.B.F.M., Morris, P.W., Voors, R.H.M., Lamers, H.J.G.L.M.: 1997, in *Luminous Blue Variables: Massive Stars in Transition*, ASP Conf. Series vol. 120, Eds. A. Nota & H.J.G.L.M. Lamers, p. 326-331 (1997)
Weis K., Duschl W.J., Bomans D.J., et al., 1997, A&A 320, 568
Westerlund B.E., 1987, A&AS 70, 311
Zickgraf F.-J, Wolf B., Leitherer C., et al. 1986, A&A 163, 119

NEW SPECTROSCOPIC DATA OF LUMINOUS PECULIAR B-TYPE STARS

F. X. DE ARAÚJO, D. F. LOPES AND M. A. MACHADO
CNPq-Observatório Nacional
Rua Gen. José Cristino, 77 - Rio de Janeiro - Brazil

1. Introduction

Since a long our group have been analysing the physical properties of some peculiar southern Be stars (see e.g. Freitas Pacheco et al., 1985; Lopes et al., 1992; Araújo and Freitas Pacheco, 1994). These stars share several common characteristics as an optical spectra dominated by Balmer and FeII emission lines (often displaying P Cygni profiles) and absorption lines of FeII and other low ionization species in the UV. Besides, prominent emissions are usually seen in the near IR and some of them are IRAS sources. Recently we have started a new study program on them, with the following goals:
i-) to obtain new, high-quality data of such objects
ii-) to perform the fitting of some lines (H_α, H_β, HeI λ 5876 Å) employing a numerical code developped by one of us (M.A.M.)
iii-) to estimate mass-loss rates and velocity fields, on the basis of the line profile fittings.
iv-) to analyse the radiative force acting on them, using a hydrodynamical code (Araújo, 1995).
v-) to try to clarify the evolutionary status of them and their relationship with other stellar classes (for instance LBVs).

In the present work we restrict ourselves to describing qualitatively the data. Only preliminary results of the modelling will be commented here.

2. The data

We have employed a CCD detector and a Coudé spectrograph (dispersion $\approx$ 18 Å/mm) attached to the 1.6m telescope of Laboratório Nacional de Astrofísica, Brazil. The spectral range of each exposure is about 400 Å and we have choosed (at least) three central wavelengths: λ 5015 Å, λ 5725 Å,

A. M. Hubert and C. Jaschek (eds.), B[e] Stars, 127–129.

and λ 6560 Å. All spectra have a signal to noise ratio S/N $\geq$ 50. In May, 1996 the following objects were observed: HD 87643, HD 89249, HR Car, GG Car, CPD-52° 9243, HR 326823 and HR 327083. (We are aware that HR Car is usually classified as a LBV star). In order to begin a systematic study on their variability most of them have been re-observed in June, 1997 but these data are not included here.

Let us comment briefly now the most conspicuous features. Emission is always present in H_α and H_β. In most of times an absorption is also clear, showing a P Cygni type profile or a central absorption feature. FeII lines (m42: $\lambda\lambda$ 4924, 5018, 5169 Å) are seen as pure emissions or P Cygni profiles. HeI λ 5876 Å and λ 6678 Å are present in the majority of cases but their profiles range from (almost) pure absorptions to P Cygni ones.

HD 326823 is a special case. Their HeI lines are indeed strong in emission (for instance HeI λ 6678 Å has a peak intensity which is half of H_α). It has already been suggested that this object is highly evolved and it is probably entering in a WN phase. Our data seem to confirm such a scenario.

3. The Balmer line fittings

Our numerical method considers a spherically symmetric and stationary outflow. The mass-loss rate is an entry parameter. The expansion velocity law and the temperature profile throughout the envelope must be given in advance. Radiative transfer is solved within the context of Sobolev approximation. It is considered simultaneously with the statistical equilibrium equations for H and He. HI and HeII atomic models consist of 10 levels and 45 transitions while HeI atomic model takes into account 17 levels and 52 transitions. A preliminary fitting to H_α and H_β profiles of HD 87643 leads to the following parameters: $L = 10^5\, L_\odot$; $T_* = 16000\, K$; $\dot{M} = 5\, 10^5\, M_\odot yr^{-1}$ and $v_\infty = 1500\, kms^{-1}$. A detailed analysis of this and other stars will be presented soon.

References

Araújo, F.X. de (1995) The equatorial plane of Be stars: an outflow driven by optically thin lines?, *Astronomy and Astrophysics*, **298**, pp. 179-186

Araújo, F.X. de and Freitas Pacheco, J.A. de (1994) The massive winds of luminous peculiar B-type stars, *Astrophysics and Space Science*, **219**, pp. 267-273

Freitas Pacheco, J.A. de, Lopes, D.F., Landaberry, S.C. and Selvelli, P.L. (1985) HD 87643: a B[e] star with a cold wind, *Astronomy and Astrophysics*, **152**, pp. 101-106

Lopes, D.F., Damineli Neto, A. and Freitas Pacheco, J.A. de (1992) A spectroscopy study of luminous peculiar B-type stars, *Astronomy and Astrophysics*, **261**, pp. 482-492

Discussion

Waters: I am surprised by the combination of spectral type (B 5) and terminal velocity (1400-1800 $km.s^{-1}$). This is quite unusual ! Could it be that the star is in fact much hotter and that the continuum is highly veiled due to free-bound emission from the ionised wind?

Araujo: Yes, this could perhaps be the situation, but it would require quite a large free-bound opacity. At the moment we have no computations which could prove or discard this idea.

Lanz: Superionization is not the only signature of a two-component wind model. This is also strongly supported by the presence of broad P Cygni lines together with narrow emission lines in the spectrum of the star.

Araujo: Well, I agree that any clear observational evidence of two quite different velocity fields points in favor of a two component model. But I did like to emphasize that the presence of superionized lines indicates not only two distinct expansion velocities but also ionisation stages and in my opinion this is a a stronger evidence of two regions in the envelope.

Israelian: From where did you obtain the force multiplier parameters α, κ, δ?

Araujo: from the well known Abbotts (1982) table.

Owocki: Could you show which spectral features provide the clearest evidence of a wind with velocity of order 1000 $km.s^{-1}$ terminal. Could you also clarify what spectral characteristic leads you to infer such a large value of the β velocity law exponent. Would not a more conventional explanation be that this is a two-component model with high-speed wind at pole and a slow dense disk outflow at equator?

Araujo: The edge velocity given by the absorption component of the P Cyg type profile is almost 1400 $km.s^{-1}$. The large β value is needed in order to fit the peak of H_α line, but it is true that it was obtained in the context of a spherically symmetric model. If a kind of equatorial disk is assumed, with a slow expansion, it will be possible to have a much more rapid acceleration in the polar directions.

DIELECTRONIC RECOMBINATION OF Mg II. INFLUENCE IN THE INFRARED CONTINUUM OF B[e] STARS.

ALICIA CRUZADO AND ADELA E. RINGUELET
Observatorio Astronómico de La Plata
Paseo del Bosque s/n, 1900-La Plata, Argentina

AND

HÉCTOR O. DI ROCCO
Instituto de Física de Arroyo Seco
Pinto 399, 7000-Tandil, Argentina

Abstract.
We investigated the dielectronic recombination of MgII in low density and low temperature media; we concluded that it may contribute to the IR excess in Be stars.

1. Introduction

The IR spectral region of Be stars has been classified by Allen (1972) into X, F and D classes, according to its similarity to normal stellar continua, free-free emission continuum and dust emission respectively.

Since the early investigation by Gehrz et al (1974) it was recognized that free-free emission could be a contributor to the IR excess. The results by Waters (1986) and Waters, Coté and Lamers (1987) were consistent with this hypothesis; Vazquez and Cidale (1993) obtained a rigorous solution of the transfer problem and were able to reproduce F type excesses.

The largest IR excesses are the ones of type D and, eventhough a large number of cases is not available, they seem to be related to the [Be] phenomenon. It was suggested by Pecker (1962,1963) that these continua were originated by dust thermal reemission. These IR excesses have the peculiarity of developping one or two bumps which are identified with Black-body curves.

A. M. Hubert and C. Jaschek (eds.), B[e] Stars, 131–133.

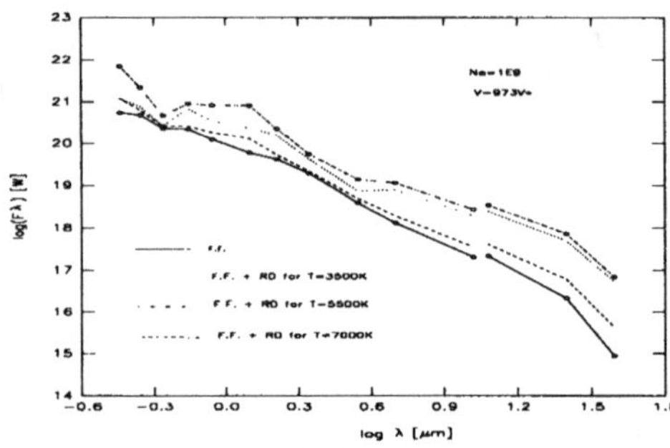

Figure 1. Flux distribution against wavelength for three different values of the electron temperature: 3500 K, 5500 K and 7000 K.

2. A suggested new mechanism

After our investigation of the dielectronic recombination of MgII into autoionizing levels and its subsequent cascade to lower energy states, we found that two bumps are obtained in the region $1\mu \leq \lambda \leq 50\mu$ and that the intensity of these bumps depends on the relevance of different atomic configurations fact which, in turn, depends on density and temperature. Our calculations yield, for normal stellar abundances and standard conditions in the envelope, an excess of 2.5 magnitudes in the L band if the emitting volume is contained in $3 \leq r \leq 10$ stellar radii. In Figure 1 illustrate some of results. Different curves correspond to different temperatures (fixed density). The emission was calculated considering 11 autoionizing configurations and all bound levels with principal quantum number $n \leq 12$ and azimuthal quantum number $l \leq 3$.

Our conclusions are,

1. the relative significance of the second bump is due, mainly, to density variations,
2. there is a fixed temperature at which the intensity of the bumps is a maximum.

The detailed calculations are to be published elsewhere.

References

Allen, D. A. 1973, Mon. Not. R. astr. Soc., 161, 145
Gehrz, R. D., Hackwell, J. A., Jones, T. W. 1974, ApJ, 191, 675
Vazquez, A. C., Cidale, L. S., Ringuelet A. E. 1993, 419, 286
Waters, L. B. F. M. 1986, A&A, 162, 121
Waters, L. B. F. M., Coté, J., Lamers, H. J. G. L. M. 1987, 185, 206

Discussion

Houziaux: I am wondering if the [OI] lines at 6300, 6364 and 5577 Å did not appear in the spectrum of HD 87643. It would then be the brightest object with such lines in its spectrum.

Cruzado: If the comment refers to the fact that the star (HD 87643) is bright in the 0.55-0.65 μ region (V filter) and [OI] is not observed, it is not probable that DR of MgII mechanism would provide the emission in that particular region, since our calculated curves show a depression at $\lambda = 0.5\ \mu$. The maximum emission in the V filter is obtained with the temperature of the envelope approximately 6.5. 10^3 K.

SPECTROSCOPY AND PHOTOMETRY OF THE B[e]–STAR CANDIDATES AS 78 AND LS II +22°8

SHEJKINA, TATIANA A.
Fesenkov Astrophysical Institute of the Kazakhstan National Academy of Sciences, Almaty, 480068, Kazakhstan

AND

MIROSHNICHENKO ANATOLY S.
Central Astronomical Observatory of the Russian Academy of Sciences at Pulkovo, Saint-Petersburg, 196140, Russia

Two poorly–studied emission–line stars AS 78 ($V \sim 11\overset{m}{.}2, B-V \sim 0\overset{m}{.}7$) and LS II +22°8 ($V \sim 11\overset{m}{.}9, B-V \sim 0\overset{m}{.}6$) were identified with IRAS sources 03549+5602 and 19381+2224 respectively by Dong & Hu (1991) among other sources having very strong infrared excesses, V-[25] $> 8^m$. Only one *UBV* photometric observation has been published (Haug 1970) for AS 78 which was discovered by Miller & Merrill (1951) as a Be star. A spectral type of B9 Ia or B2–5 V was estimated for LS II +22°8 by Pesch (1963) from the *UBV* photometry while near–IR data ($K = 9\overset{m}{.}6$, Allen 1973) show a very weak excess in this region. Here we present our recent observational results for these stars.

In 1993–96 we obtained their medium–resolution spectroscopy at the 6–meter Russian telescope and photometry in the UBVRI–bands with two 1–meter telescopes of the Fesenkov Astrophysical Institute. Both objects show photometric variability of the order of $0\overset{m}{.}2$ and strong $H\alpha$ emission lines with equivalent widths of 40 (LS) and 75 Å (AS).

We estimated spectral type of AS 78 as B5 from its *UBV* colors. Its spectrum is enriched with weak emission lines of Fe II and Ti II, the Balmer lines have P Cyg profiles with emission components seen together with broad photospheric lines which is similar to that observed in FS CMa (Israelian et al. 1996). In order to determine physical parameters of the star and its wind we modeled the Balmer line profiles using the Sobolev method to calculate the level populations combined with the exact solution of the radiation transfer problem in a spherical envelope (Pogodin 1986). A β–velocity law was used to describe the wind kinematics. The best results were

A. M. Hubert and C. Jaschek (eds.), B[e] Stars, 135–136.

obtained for the following parameter values: $T_* = 16000K$, $R_* = 10R\odot$, $log\, g = 3.5$, $\log L_{bol}/L\odot = 3.75$, $v_\infty = 450$ km s^{-1}, $v_0/v_\infty = 0.0025$, $\dot{M} = 3\,10^{-6} M \odot yr^{-1}$, β=4, where v_0 and v_∞ are the velocity at the base of the wind and terminal velocity respectively.

We obtained the K–band photometry of AS 78 on 1996 December 21 and detected a significant IR–excess ($K = 6^m\!.4 \pm 0^m\!.2$). Near– and far–IR excesses observed in its spectral energy distribution (SED) allow to assume that the object is surrounded by a dusty envelope. A model to fit the observational SED was computed with a code by Ivezić, Nenkova, & Elitzur (1997) solving radiative transfer problem in a spherical dusty envelope. The fitting results are described by Ivezić et al. (this volume). All the parameters derived from both the SED and line profile modeling give a distance estimate toward AS 78 of nearly 3 kpc. Having a luminosity close to the ZAMS, the star could be either Herbig Ae/Be star or an object in the horizontal branch evolving to the planetary nebula stage.

Fluxes measured for a possible IRAS counterpart of LS II +22°8 (S_{12} = 10.43 Jy, S_{25} = 7.10 Jy, S_{60} = 3.64 Jy) imply the presence of an additional source of IR radiation. Thé et al. (1994) pointed out that the optical position of LS II +22°8 and the IRAS source were ofsetted by 54″. The LRS spectrum of the IRAS 19381+2224 shows a weak emission feature at 11.3 μm which is probably due to SiC dust particles usually observed in the spectra of carbon stars. Therefore, association of this IRAS source with LS II +22°8 is doubtful.

The most prominent features in the LS II +22°8 spectrum are a double–peaked $H\alpha$ emission and Fe II absorption lines. The $H\beta$ line is filled in with emission, $H\gamma$ and $H\delta$ show weak signs of emission. Absorption line profiles of both H I and He I show the best agreement with B9–A0 Ia types. Strong Fe II absorptions are usually weak in such supergiants while getting stronger with luminosity increase and stellar temperature decrease. If we suggest that LS II +22°8 has an absolute magnitude of at least -5^m it would be at a distance of 10 kpc from the Sun. Thus, our study shows that LS II +22°8 is more likely a luminous star with a strong mass loss.

References

Allen, D.A. (1973) *MNRAS*, **161**, 145
Dong, Y.S., Hu, J.Y. (1991) *Chin. A&A*, **15**, 275
Haug, U. (1970) *A&AS*, **1**, 35
Israelian, G., et al. (1996) *A&A*, **311**, 643
Ivezić, Ž., Nenkova, M., & Elitzur, M.: 1997, *Univ. of Kentucky, Internal Report*
Miller, W.C. & Merrill, P.W. (1951) *ApJ*, **113**, 624
Pesch, P. (1963) *ApJ*, **137**, 547
Pogodin, M.A. (1986) *Astrofizika*, **24**, 491
Thé, P.S., de Winter, D., & Pérez, M.R. (1994) *A&AS*, **104**, 315

ANALYSIS OF THE EMISSION LINE SPECTRUM OF V380 ORI AND HD45677

G. MURATORIO[1], C. ROSSI[2], M. FRIEDJUNG[3], R. VIOTTI[4]
1. Observatoire de Marseille, 2 Place Le Verrier, F-13248 Marseille Cedex 04, France
2. Istituto Astronomico dell'Università "La Sapienza", Via G.M. Lancisi 29, I-00161 Roma, Italy
3. Institut d'Astrophysique (CNRS), 98 bis Boulevard Arago, F-75014 Paris, France
4. Istituto di Astrofisica Spaziale (CNR), Via E. Fermi 21, I-00044 Frascati RM, Italy

Abstract. The ionized metal emission lines in the Herbig Ae/Be star V380 Ori and in the B[e] star HD45677 (FS CMa) have been analysed with the SAC method. We found that probably all the emission lines are optically thick. The analysis of V380 Ori indicates that the principal components of the Fe II emission lines are very probably formed in a rotating disk.

The self absorption curve (SAC) method described by Friedjung (this conference) has been applied to the Herbig Ae/Be star V380 Ori (Rossi et al. 1997, A&A submitted) and to the B[e] star HD45677 (FS CMa, paper in preparation).

In V380 Ori the strongest Fe II emission lines have a complex profile with a principal central emission and a blueshifted (-140 km s^{-1}) component. The SAC of the Fe II principal components is shown in Fig. 1. As can be clearly seen also examining individual multiplets, the SAC is almost linear over three decades in τ with a slope of -0.47, indicating that *all the optical Fe II emission lines are optically thick.* The populations of the odd upper and even lower levels of the Fe II lines approximate two different Boltzmann laws with excitation temperatures of 3800 and 2900$^\circ K$, respectively. The lower limit to the hydrogen column density, assuming cosmic abundances and the same even level Boltzmann law extending to 0 eV, is 2 10^{26} cm^{-2}. This assumes that the width of each line in velocity units *along a line of sight* v_o is thermal, leading to more consistent results than

A. M. Hubert and C. Jaschek (eds.), B[e] Stars, 137–139.

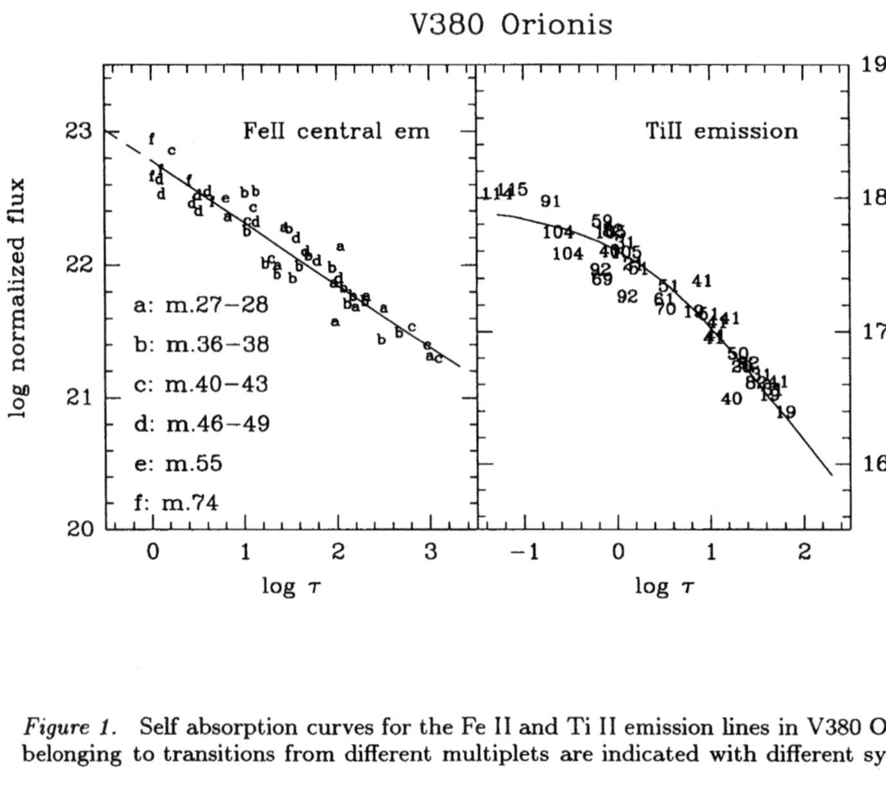

Figure 1. Self absorption curves for the Fe II and Ti II emission lines in V380 Ori. Lines belonging to transitions from different multiplets are indicated with different symbols.

if v_o is assumed equal to the full *observed line width.* Our assumption is in agreement with line formation in *a thin rotating disk.* The minimum Fe II emitting region characteristic radius R_o of 1.2 10^{13} cm, is then much larger than that of the stellar photosphere. The observed Ti II emission lines do not give such strong constraints concerning the line emission region, while the blueshifted emission components of Fe II appear also to be optically thick.

In the case of HD45677 narrow and broad components are seen for the permitted Fe II emission lines, while the forbidden lines are narrow. Preliminary analysis of the permitted lines indicate that they are probably optically thick with most excitation temperatures higher than those found in V380 Ori.

Discussion

Araujo: Could you say which is the main excitation mechanism of the FeII lines in each case (B[e] star/Herbig Ae/Be)?
Friedjung: We can have reversal mechanisms; the SAC method can be applied without knowing the details. When the lines are optically very thick, collisional excitation becomes more important relative to radiative excitation.
Cassinelli: Could the Fe II be formed in the region beyond the dust formation zone?
Friedjung: I am not 100 % certain though my impression is that Fe II emission line formation probably occurs at rather smaller radii than the dust.

LONG-TERM SPECTROPHOTOMETRIC BEHAVIOUR OF B[e] STARS

A. MOUJTAHID[1], J. ZOREC[2], A.M. HUBERT[1]
[1] *Observatoire de Paris-Meudon, F-92195 Meudon Principal. Cedex*
[2] *Institut d'Astrophysique-*CNRS, *98 bis, bd. Arago, F-75014 Paris*

1. Aim, data and method

Classic Be stars show characteristic linear relations between luminosity excess and intrinsic reddening in the visible range against the total Balmer discontinuity (BD) (6). These relations differ according to whether the star is in a "Be" or a "Be-shell" phase, so that the limit between the two spectrophotometric behaviours (SPB) may be used to infer a rough spectral type of the underlying star. As circumstellar dust absorption cannot strongly affect BD determination, it is worth trying to identify some kind of a break between Be-like and Be-shell-like behaviour in B[e] stars produced by the gaseous circumstellar envelope, for their spectral type is difficult to estimate otherwise.

As for classic Be stars, we determine the colour gradient Φ_{rb} [μm] and the BD, $\mathcal{D}$ [dex] [definitions of parameters follow the BCD system (4)] using UBV and *uvby* photometric data [(1), (2), (5), (7)]. Transformations from photometric colour indices to spectrophotometric parameters Φ_{rb} and $\mathcal{D}$ are given and discussed in (6). The studied B[e] stars and observation periods are: MWC 314 (1990.7-94.7), MWC 623 (1989.7-91.9), MWC 342 (1990.-93.), MWC 84 (1989.8-94.8), HD 51585 (1968.-92.) and HD 45677 (1973.-94.).

2. Results obtained

The time interval covered by observations is rather short, so we probably miss many aspects of the stellar SPB. As B[e] stars show strong and variable line emission, photometric indices may produce spurious SPB wich introduce heavy scatter in the studied spectrophotometric relations. Un-

A. M. Hubert and C. Jaschek (eds.), B[e] Stars, 141–143.

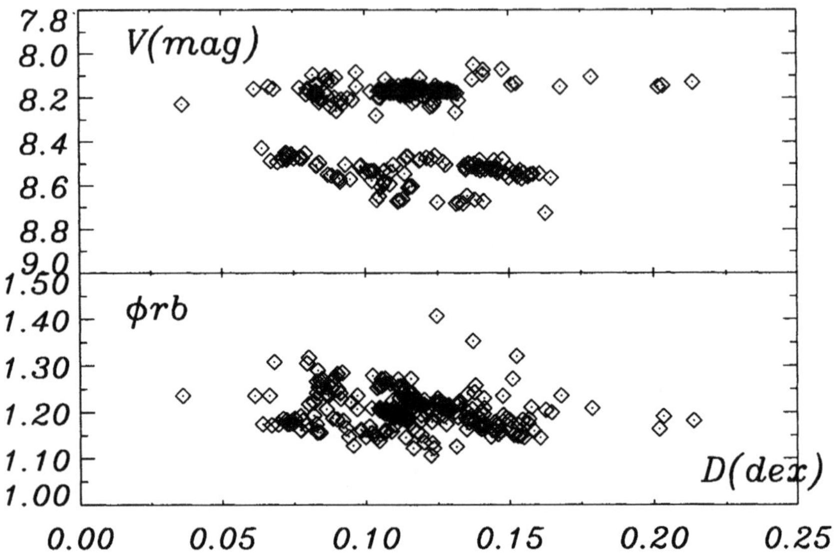

Figure 1. V magnitude and Φ_{rb} gradient against the Balmer discontinuity $\mathcal{D}$ of HD 45677

fortunately, due to such uncertainties in the values of U, V, Φ_{rb} and $\mathcal{D}$, we could not obtain the stellar BD of the studied B[e] stars. However, there is a vague indication for a stellar BD in MWC 84 and HD 51585. In all studied stars we noted that the magnitude V remains almost constant, although U and $\mathcal{D}$ have a large amplitude of variation. We also note that $\Phi_{rb} \simeq$ constant. It is also worth noting the peculiar behaviour of HD 45677. Two different episodes of V $\sim$ constant are seen and for each episode also $\Phi_{rb} \simeq$ constant (see Fig. 1). Jumps in the U, B and V magnitudes are the same, as there would be a grey absorption effect. Such a phenomenon could be explained by an effective screening effect of the underlying star produced by a clumpy CE, where the optical depth of clumps is high (3).

References

Arkhipova, V.P., Ikonnikova, N.P. (1992), *Sov. Astron. Letters*, **18**, p.418
Bergner, Y.K., Miroshnichenko, A.S., Yudin, R.V. et al. (1995), *A&AS*, **112**, p.221
Boissé, P. (1990), *A&A*, **228**, p.483
Chalonge, D., Divan, L. (1952), *Ann. Ap.*, **15**, p.201
Feinstein, A., Garnier, R., Vogt, N. et al. (1976), *A&A*, **51**, p.269
Moujtahid, A., Zorec, J. Hubert, A.M. et al. (1997), *A&AS* (in press)
de Winter, D., van den Ancker, M.E., Perez, M.R. et al. (1996), *A&AS*, **119**, p.1

Discussion

Viotti: Your observations of B[e] stars seem anyhow to indicate the presence of a variable Balmer continuum. Is this an indication of a Balmer continuum emission?
Moujtahid: It may be as well emission or absorption. Many B[e] stars have strong second Balmer discontinuity in emission.
Hubert: What is the error bar of the determination of D_* from your graphs on B[e] stars (and also Be stars)?
Moujtahid: We used the UBV photometric system to determine the Balmer discontinuity. The error on D_* are, hence, of the order of 0.03 dex.

PHOTOMETRY, POLARIMETRY, AND INTERFEROMETRY OF B[e] STARS. VARIABILITY

MIROSHNICHENKO ANATOLY S.
Central Astronomical Observatory of the Russian Academy of Sciences at Pulkovo, Saint-Petersburg, 196140, Russia

1. INTRODUCTION

Twenty years ago Allen & Swings (1976) found a group of early-type emission-line stars having infrared (IR) excess at 2–3 μm too large to be explained by circumstellar gas emission only. Most of them show forbidden lines what has caused a popular name of these objects – B[e] stars. The criteria allowing to distinguish 65 such objects were as follows:

1. Spectral type – mainly B
2. Presence of emission lines
3. Excess of near-infrared radiation

Classical Be stars showing smaller IR-excesses due to free-bound and free-free radiation as well as WR stars and Herbig Ae/Be stars (HAEBEs) were not included to the list although the latter display IR-excesses due to circumstellar dust like B[e] stars. Allen & Swings introduced three groups of B[e] stars: group 1 (17 stars) with weak emission spectra of Balmer lines, Fe II and, occasionally O I, group 2 where forbidden lines of low-ionized species are dominant (26 stars), and group 3 with high ionization states such as [O III], [Ne III] (22 stars). They proposed three different scenarios to explain formation of the dusty envelopes:

1. Formation of planetary nebulae as a single event
2. Interaction of an OB star with a late-type companion
3. Direct ejection from a particularly massive Be or Oe star

Carlson & Henize (1979) attracted attention to a group of 20 southern early-type emission-line stars which imply a significant degree of mass ejection. This group can be considered as a first supplement to the sample of galactic B[e] stars because of the two facts. Most of them turned out to have near-IR excesses and could, therefore, be classified as B[e] stars.

A. M. Hubert and C. Jaschek (eds.), B[e] Stars, 145–164.

Four stars are common in these two lists: CPD -52°9243, HD 87643, HD 167362, and GG Car. Two objects from the Carlson & Henize's list, He3–759 and He3–1213, have not been observed in the near–IR. Thus, taking into account the rest of them, 79 objects could form a group of galactic B[e] stars by the end of 70s.

During follow up studies 47 % of them have been recognized as the objects at well–understood evolutionary stages: pre–main–sequence HAEBEs (7), LBVs (3), PNe (16), symbiotic systems (8), eclipsing binaries (2: RY Sct, GG Car). Other objects fall into the several categories:

1. Supergiants similar to those in the Magellanic Clouds found by Zickgraf et al. (1986) – 6 stars (3 Pup, MWC 300, MWC 349, HD 87643, HD 316285, HD 326823)
2. Probable binary systems consisting of a hot and a cool star – 6 (MWC 84, CPD-52°9243, MWC 623, CPD-57°2874, HD 89249, HD 327083)
3. Well–observed objects with puzzling nature – 4 (FS CMa, HD 50138, MWC 342, CD -42°11721)
4. Poorly-studied stars with uncertain state – 27 stars

121 mostly unstudied objects with large IR–excesses at the IRAS wavelengths ($V - [25] > 8^m$) have been found by Dong & Hu (1991) by means of cross–identification of the objects from the Wackerling (1970) catalog of galactic early–type emission–line stars and IRAS sources. Eleven of them belong to the Allen & Swings' list, 4 – to that of Carlson & Henize, 17 have been recognized as HAEBEs candidates, carbon stars, X–ray binaries, WR stars, LBV candidates, symbiotic systems, PNe candidates. The rest 90 objects wait for their investigators and can be a potential reservoir for new candidates to the B[e] group.

Recently Thé et al. (1994) analysed observational characteristics of the early–type stars with evidence for dusty envelopes trying to separate those at the pre–main–sequence evolutionary stage. They found no unique observational data set allowing to do this and compiled a new catalog of "Herbig Ae/Be stellar group" containing 287 currently known candidates to HAEBEs, B[e] stars, and, unstudied objects listed by Dong & Hu.

In the present paper pre–main–sequence star candidates were not included except those with controversial evolutionary state. Its purpose is to review information obtained so far for the above four categories of B[e] stars, as well as for some recently studied objects from the Dong & Hu's list which can be considered as B[e] stars candidates.

Photometric and polarimetric observations of B[e] stars have been obtained so far mostly at an occasional basis. There are only a few programs in which long–term photometric observations of a number of B[e] stars have been carried out: Kilkenny et al. (1985), Manfroid et al. (1991), Sterken et al. (1993, 1995), Thé et al. (1994), Bergner et al. (1995). An ongoing program including classical B[e] stars and recently found candidates is described by Kuratov et al. (this volume). Polarimetric observations were carried out by Barbier & Swings (1982), Zickgraf & Schulte–Ladbeck (1989), and reported in the papers quoted in Table 3. A lot of results are still unpublished, and I am very grateful to the owners of these observations for sending them to me for the preparation of this review.

2. PHOTOMETRIC PROPERTIES

According to a number of photometric observations obtained we can divide B[e] stars into three groups: those with light–curves (Table 1), those with spectral energy distributions (SEDs) known in a wide spectral range (from UV to far–IR) but rarely observed (Table 2), and those with data in quite a narrow spectral range.

2.1. OBJECTS WITH LIGHT CURVES

TABLE 1. Objects with numerous observations

MWC	Name	IRAS	Nature	Number of obs.		ΔV
				Optical	Near–IR	
84	CI Cam	04156+5552	Binary ?	49	25	0.3
142	HD 45677 FS CMa	06259−1301	HAEBE ?	~400	38	2.0
158	HD 50138	06491−0654	HAEBE ?	155	16	0.4
198	HD 87643	10028−5825	SG ?	202	11	1.0
300		18267−0606	SG/HAEBE?	380	9	0.4
623	V2028 Cyg	19545+3058	Binary	31	27	0.2
342	V1972 Cyg	20212+3920	Binary ?	570	81	0.6
349	V1478 Cyg	20310+4029	SG	25	31	0.4
930	V446 Sct	18237−0715	Binary ?	228	13	0.3

MWC 84, MWC 623, and MWC 930 have been recently added to the General Catalog of Variable Stars by Kazarovets & Samus (1997) based on the data of Bergner et al. (1995). Information on variability of HD 87643 has not been previously summarized, so that it has no designation as a variable star.

Photometric history of the best observed object, **FS CMa**, has been recently summarized by de Winter & van den Ancker (1997) for a period since the beginning of the century. Its visual brightness varied between $7^{m}.4$ and $6^{m}.9$ untill early 60s and then declined to $8^{m}.8$ in some 20 years. Since the minimum in 1981 the star continues brightening untill the present time reaching a level of $8^{m}.0$. Multicolor photoelectric photometry obtained since late 50s showed that its optical color–indices changed a little on background of strong brightness changes. A trend can hardly be seen in $B-V$ while dispersion of other color–indices veils it. Quasi–regular brightness variations with a period of $296^{d}.5$ have been found by Halbedel (1989) for a four–year data set. Our analysis of the observations obtained in 1975–1996 shows presence of significantly slower variations with a period of $\sim 1600^{d}$ (Fig.2). Their signature is also visible in the object's behavior before the decline. Variations in the J and H bands follow the optical changes even displaying some traces of the 1600^{d} cycles. Variability at longer wavelengths is not studied enough because of small number of observations. Sitko et al. (1994) comparing two SEDs obtained between 0.3 and 20 μm in 1980 and 1992 concluded that changes in the visual and mid–IR were opposite. Photometry in the 10–μm band obtained in late 60s shows that the object was at the 1992 level at that time. At the same time, mean brightness in a region between 2 and 8 μm is stable for more than 25 years. Grady et al. (1993) detected high–velocity gas moving toward the star in the IUE spectra which has also been found for a number of HAEBEs.

HD 50138, a very similar object with respect to FS CMa but having later spectral type, demonstrate the same photometric behavior. It was as faint as $6^{m}.8$ in 1982–1983 (Kilkenny et al. 1985) and since that time has been increasing visual brightness. At the moment it reached a level of $6^{m}.5$ at which it was before 80s. Optical color–indices have not changed like in FS CMa. Near–IR brightness follows the optical changes with decreasing amplitude towards longer wavelengths. This is expected for the case of formation of a dusty envelope with less optical depth than that of FS CMa. Periodicity can hardly be found because of smaller data set. Accretion phenomena have been found in the IUE spectra of the star (Grady et al. 1996) as well as in the optical region (Pogodin 1997).

A bright southern IR source, **HD 87643**, is known to have a spectrum similar to that of η Carinae and, thus, to be a candidate to B[e] supergiants (see Araujo et al., this volume). It declines almost linearly since 60s displaying variations around the mean track of the order of $0^{m}.5$ (Fig.1). The optical color–indices get redder with the brightness decline except $U-B$. The latter turned out to be bluer until $V = 9^{m}.0$ and redder when the star declined further. We found a quasi–periodicity of $925^{d}.3$ by means of Fourier analysis of its visual light curve (Fig.2). The most recent observations (Torres et al. 1995 and our data obtained at the SAAO on 1997 July 10) show that now the star is as faint as $V = 9^{m}.5$. Near–IR observations of

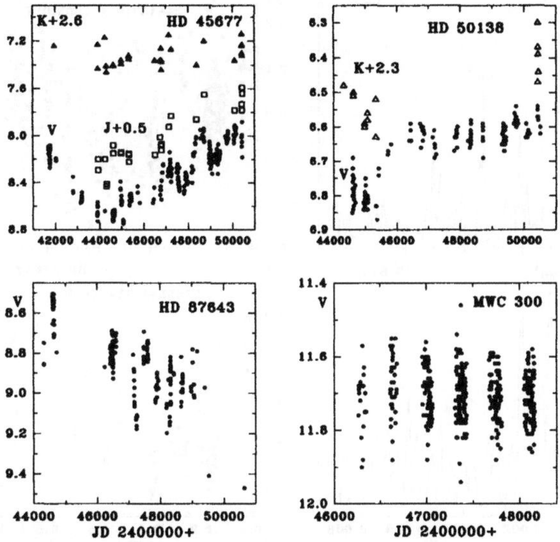

Figure 1. Light curves of the well-observed B[e] stars

the object are rare but the data show no trend in the JHK bands and 20 % increase of brightness in the M–band between two observations obtained in early 70s and in 1988. If we compare its mean SEDs of 1983 (when the IRAS data were obtained) and of 1988, we can see the same effect as that in FS CMa: decline in the optics and brightening in the mid–IR. It is very important to continue monitoring of this object.

MWC 300, an object considered as a B[e] supergiant (Wolf & Stahl 1985) or a HAEBE star (Finkenzeller & Mundt 1984), shows no brightness trend over the last 20 years and no periodical variations. An amplitude in the V–band is nearly $0\overset{m}{.}35$, $U-B$ and $B-V$ do not depend on the brightness level. A very weak reddening with the brightness decrease is noticed only in $V-R$. If its luminosity is $M_{bol} = -9\overset{m}{.}5$ (Wolf & Stahl 1985) a distance toward it should be as large as 15 kpc.

Attention to **MWC 342** has been attracted by Bergner et al. (1990) who detected a 132^d cycle most pronounced in the U–band. A cycle with a half of this period was also found in the R–band polarization. Since 1989 a group of V.S.Shevchenko obtained more than 450 $UBVR$ photometric observations of this source. Their results show the presence of the 132^d cycle only in the 1989 set. Later it disappeared. In general, the star displays gradual brightening in the optical and near–IR region (see Fig.8). Closeness of a soft X–ray source to the star motivated Miroshnichenko (1991) to suggest that it is a Be/X–ray binary.

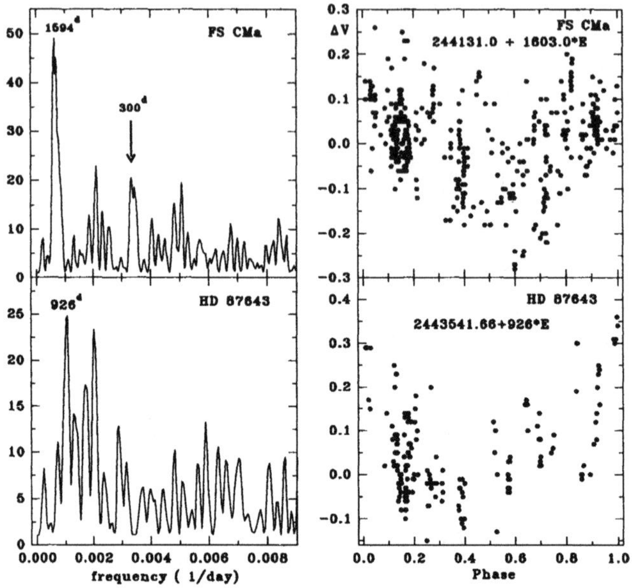

Figure 2. Power spectra of the light curves and phase curves for the most pronounced periods of FS CMa and HD 87643. 300^d variations are removed from the phase curve of FS CMa what slightly changed the best value of slow variation period.

MWC 84 and **MWC 930** have been observed by Bergner et al. (1995) and Shevchenko et al. (1993). Miroshnichenko (1995) found quasi–periodic variations in these data ($11\overset{d}{.}7$ and $58\overset{d}{.}4$ respectively) and suggested that they are binaries because of anomalous Balmer decrements (see Fig.5), features of late–type stars in their spectra, and photometric behavior. MWC 930 was not included to the Allen & Swings' list because of small near–IR color–indices, however, it turned out to have a strong far–IR excess and emission lines including those of extremely strong Fe II and can be classified as a B[e] star.

MWC 623 was first suggested to be a binary system consisting of a hot and a cool star by Arkhipova & Ipatov (1982) from a spectrophotometric study (B0 V + M1 III) and then by Zickgraf & Stahl (1989) by means of high–resolution spectroscopy (B2 V + K2 II - III). The difference in spectral type estimates was partly due to the absence of the object's optical photometry by that time. Bergner et al. (1995) obtained ~ 30 *UBVRIJHK* observations, modeled the object's SED, and derived the B2 + K7 combination for the spectral types of the companions. The MWC 623 SED in the IRAS region is similar to those of late–type stars. No regular variability has been found in the photometric data.

MWC 349 is a well–known bright IR and radio source. It most likely has been found in the photometric data.

MWC 349 is a well–known bright IR and radio source. It most likely belongs to the Cyg OB2 association which stars display a very strong interstellar reddening. Based on this, MWC 349 is considered to be a B[e] supergiant. It exhibits $\sim 0\overset{m}{.}4$ variability in the *VRIJHK*–bands which is correlated in all these bands (Bergner et al. 1995). Analysis of this small data set allows to suggest a cyclicity with a period of $\sim 1000^d$. Gottlieb & Liller (1978) studied the photographic light–curve of MWC 349 based on Harvard plates obtained between 1900 and 1980 and also found evidences for slow cyclic brightness changes.

Thus, less than 20 % of B[e] stars were observed enough to analyze their light curves. The averaged amplitude of their optical variability is nearly $0\overset{m}{.}5$, which is significantly larger than that obtained for the Magellanic Clouds B[e] supergiants. However, we should keep in mind that these stars have been observed within quite short time intervals. More than a half of them, no matter are they suspected to be binaries or not, display quasi–periodic brightness changes which nature is not understood at the moment. Two stars, FS CMa and HD 87643, display long–term high–amplitude variations of visual brightness which could be due to recent and probably recurrent dust formation events. Recently Arkhipova et al. (1996) using old photographic plates found another star, HDE 341617, showing similar photometric behavior. This approach being applied to other rarely observed B[e] stars could reveal new such objects allowing to study this process in more detail.

2.2. SPECTRAL ENERGY DISTRIBUTIONS

More than 20 other B[e] stars have been observed occasionally and, in many cases, non-simultaneously in different spectral regions. Nevertheless, their SEDs can be constructed in a wide spectral range (see Fig.6). Basic data on these stars collected from the literature as well as from unpublished sources are presented in Table 2.

Analysis of their optical color–indices shows that in color–color diagrams they locate close to the mean reddening vector for the hottest stars or even above it. This means that they certainly contain a high–temperature star, as it is seen from spectroscopy, dominating the UV and blue spectral region. However, it turns out to be difficult to separate binaries containing a hot and a cool star and single stars. For comparison VV Cep–type binaries containing an OB–dwarf and an M-supergiant are shown in Fig 3. In many cases brightness ratio of the components makes objects very similar to highly reddened hot stars. However, VV Cep stars display smaller IR–excesses and steeper flux decrease toward longer wavelengths in the IRAS

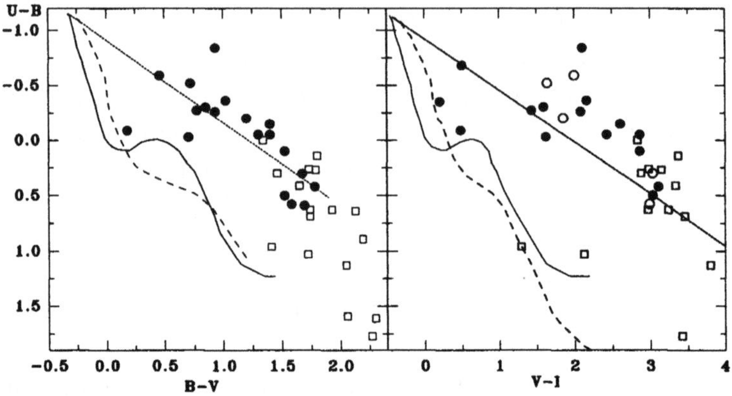

Figure 3. Optical color–color diagrams. B[e] stars are denoted by filled circles, VV Cep stars by open squares, possible B[e] binaries by open circles. Solid line represents normal color–indices of dwarfs, dashed line shows those of supergiants

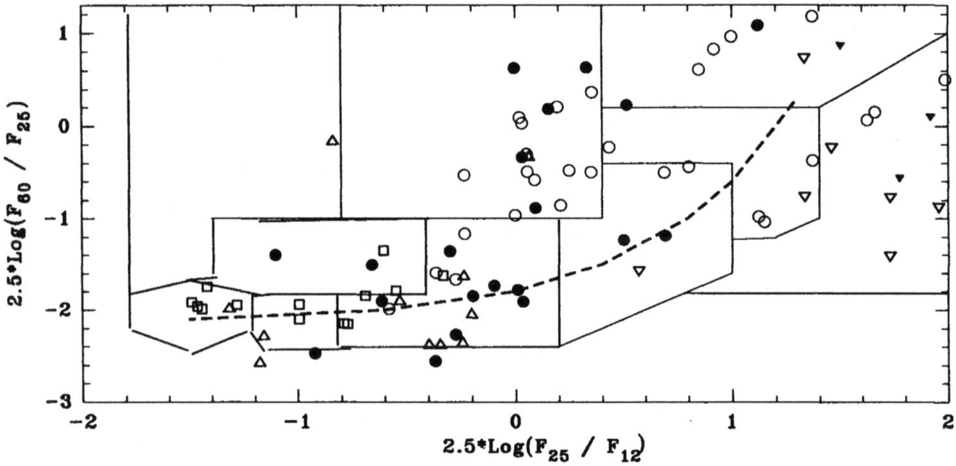

Figure 4. The IRAS color–color diagram. B[e] stars are denoted by filled circles, VV Cep stars by open squares, symbiotic systems by open up triangles, B[e] stars recognized as Planetary Nebulae by open down triangles, young PNe from the Parthasarathy & Pottasch's (1989) list by filled down triangles, HAEBEs by open circles. Dashed line represents the track for objects evolving toward PNe from van der Veen & Habing (1988). Solid lines split the diagram into regions containing different type of objects introduced by van der Veen & Habing.

As was mentioned above, several B[e] stars have been suspected to be binaries by means of spectroscopic observations (MWC 623, HD 89249, 3 Pup, MWC 84, MWC 930). Three of them locate in the cool star region in the [25] – [12] $\sim$ [60] – [25] diagram. IRAS data for HD 89249, not an

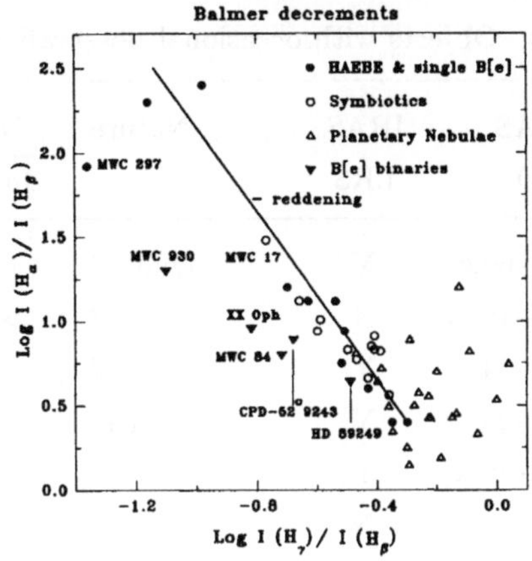

Figure 5. Balmer decrements of different types of objects with IR–excesses

IRAS PSC source, were obtained using the ADDSCAN procedure. This star, as well as MWC 930, shows an additional excess in the region of $\lambda > 25\mu m$. The SED of MWC 930 is similar to those of LBVs, but the observed features mentioned above tell in favor of the binary hypothesis (Miroshnichenko 1995). It is shown in Fig.5 that Balmer decrements could be a tool to distinguish binary stars when the cool companion dominates the red continuum of the system. A HAEBE candidate MWC 297 also having an unusual Balmer decrement was suspected to be a binary star by Miroshnichenko (1994), however recent study of its spectrum by Drew et al. (1997) did not confirm this suggestion.

Several other B[e] stars (CPD-52°9243, CPD-57°2874, HD 327083, He2–139, M1–76, and probably AS 225 which 12 μm flux has been measured with a poor quality) also fall in this region. He2–139 was recently recognized as a symbiotic system. Both CPD stars and the HD star were found to show CO first overtone emissions at 2.3 μm (McGregor et al. 1988) which are observed in the late–type star spectra. Additionally, CPD-52°9243 displays anomalous Balmer decrement. M1–76 and AS 225 are not studied enough to make definite conclusions. Thus, there are several indications that all these stars are binaries containing a cool star besides the hot one. They have smallest IR–excesses among B[e] stars, however, higher than those of VV Cep stars. Six of them were observed by IRAS with its Low–Resolution Spectrometer. Only 3 Pup, which has a companion much fainter than the primary, shows silicate features at 10 and 18 μm in emission, other spectra are quite noisy and flat.

Recently we found an object exhibiting an intermediate IR–excess which is larger than those of possible B[e] binaries but still smaller than those of

TABLE 2. Objects with occasional observations

Name	IRAS ID	IRAS LRS	Nature	Number of obs. Optical	Number of obs. Near-IR
MWC 17	01441+6026	Y	Symb. ?	5	12
HD 51585	06556+1623	N	PN ?	>30	3
CD-24°5721	07370−2438	Y	?	5	2
3 Pup	07418−2850	Y	Asg.	>10	3
Ve 2-27	08502−4606	Y	?	2	3
Hen 230	–	N		3	2
Hen 373	10082−5647	Y	HAEBE ?	1	2
Hen 664	11289−6333	N	?	1	1
Hen 938	13491−6318	Y	HAEBE ?	1	2
CPD-52°9243	16031−5255	Y	SG./Bin?	10	5
Hen 1191	16235−4832	Y	Symb./PMS?	8	9
AS 225	17175−3757	Y	?	3	1
MWC 922	18184−1302	Y	?	2	9
MWC 939	18313−1738	Y	?	12	3
CPD-57°2874	10136−5736	Y	Bin. ?	7	5
HD 89249	–	N	Bin.	6	2
CD-42°11721	16555−4237	Y	HAEBE?	10	7
HD 326823	17033−4232	N	Pre-WR?	15	3
HD 327083	17117−4016	Y	Bin.?	6	3
HD 316285	17450−2759	Y	SG.	11	6
MWC 645	21516+5245	Y	?	1	2
AS 78	03549+5602	N	?	32	1
MWC 657	22407+6008	N	?	10	2

VV Cep stars. Six of them were observed by IRAS with its Low–Resolution Spectrometer. Only 3 Pup, which has a companion much fainter than the primary, shows silicate features at 10 and 18 μm in emission, other spectra are quite noisy and flat.

Recently we found an object exhibiting an intermediate IR–excess which is larger than those of possible B[e] binaries but still smaller than those of other B[e] stars. It is MWC 657 distinguished by Dong & Hu. We obtained its $UBVRIK$ photometry and modeled its SED in the range of 0.3 – 60 μm (Miroshnichenko et al. 1997) under assumption that it is an early–B star

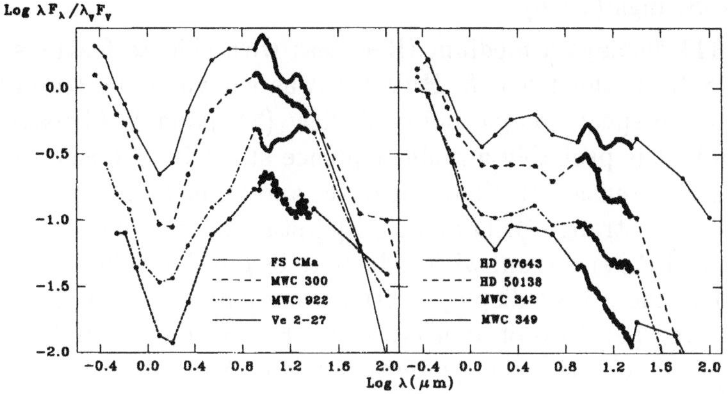

Figure 6. SEDs of B[e] stars with most strong IR–excesses.

Two high–luminosity stars, HDE 316285 and HDE 326823, display small IR–excesses in a region of $\lambda \leq 12\mu$m which can be explained by free–free radiation of gaseous envelopes. However, ADDSCAN data show that both sources appear extended at the 25–100 μm IRAS bands. This could be due to radiation from dusty envelopes ejected from the stars in the past. Another suggested supergiant, MWC 349, shows IR–excess as strong as that of MWC 657, but almost featureless LRS spectrum probably indicating a deficit of silicates in its surrounding dust. It also appears extended at 25–100 μm. The detected cyclic variations do not exclude presence of a companion. MWC 349 has a higher temperature than HDE 316285 and HDE 326823, and could be simply younger forming its envelope more recently.

A group of other B[e] stars display very strong IR–excesses. It includes HD 45677, Ve2–27, MWC 300, MWC 922, and MWC 939. The former two display clear silicate feature in emission, MWC 300 has a flat LRS spectrum, while that of MWC 922 raises towards longer wavelengths. Quite a steep decrease of the SEDs of HD 45677 and MWC 300 points out to small extension of their dusty envelopes. They both are suggested to be pre–main–sequence stars although they are isolated from star forming regions. Ve2–27 and MWC 922 appearing slightly extended at 100 μm could have larger envelopes. They are rather faint in the optical region and detailed spectroscopy is needed to make a next step in their understanding.

Two other objects, MWC 342 and HD 50138, show almost the same decrease in their far–IR SEDs but smaller excess values which can be explained by smaller dust optical depths. Optical spectrum of MWC 939 was found to be similar to that of η Carinae by Arkhipova (1975). 12 *UBV* observations obtained at the Crimean Observatory by her group in 1988 –

1993 are very close to each other as well as to the optical brightness given by Allen & Swings (1976).

HD 87643 displays a medium IR–excess with silicate features in emission. AS 78 from the Dong & Hu's list was shown to be similar to HD 45677 from the spectroscopic point of view (Shejkina & Miroshnichenko, this volume). It is probably a main-sequence star with a dusty envelope of a rather small extension (ratio of the outer and inner radii is ≤ 1000). He3–1191 has a very strange SED among B[e] stars containing a strong 3–μm bump and an LRS spectrum with a deep at 18 μm. De Winter et al. (1994) reported that the object displays very strong emission–line spectrum and have stellar temperature of at least 30000 K. They suggested that it is a pre–main–sequence star or a symbiotic system. More detailed information about B[e] SED modeling can be found in Ivezić et al. (this volume).

Other stars from Table 2 are poorly–studied. We certainly have to extend the sample of B[e] stars in order to make statistical analysis of the observed and physical characteristics possible. At the moment none of the objects of Tables 1 and 2 is clearly understood except several possible binaries and supergiants, for which some arguments in favor of this interpretation have been obtained. Nevertheless, evolutionary state of the latters is not well–determined. Another group with steep decrease of the IR–brightness probably formed their circumstellar envelopes quite recently. Perhaps they belong to the horizontal branch and are very young PNe. Otherwise they could pass an evolutionary stage near the main-sequence or after it (young B[e] supergiants) because HAEBEs usually loose their near–IR excess first. The descendants of the lower mass HAEBEs, β Pictoris stars, retain their far–IR excesses for a long time already being at the main-sequence. At the same time we do not know much about evolution of more massive HAEBEs near the main-sequence, nor about evolution of massive stars just after leaving it. Not only distances towards B[e] stars are uncertain but stellar effective temperatures are also poorly known because of the photospheric line veiling by circumstellar emission. These objects, as well as others having complex structure, require comprehensive studies using all possible observational methods.

3. POLARIMETRY

Polarimetric observations have been obtained for nearly 20 B[e] stars (Table 3). It is seen that all these stars display significant degree of optical polarization. There is a correlation between the averaged polarization (p_V) and $V - L$ color–excess due to the circumstellar envelope (E_{V-L}) found by Yudin (1988) for both HAEBEs and T Tau stars. This relation was explained by him as a consequence of the dusty origin of the polarization.

Comparing HAEBEs and B[e] stars in the log $(p_V) \sim E_{V-L}$ one can say that B[e] stars have, in general, lower polarization degree and higher excesses. At the same time, B[e] stars exhibit a stronger correlation between polarization and overall extinction A_V than HAEBEs (Fig.7).

TABLE 3. Polarimetric data on B[e] stars

Name	p_V	Θ_V, °	E_{V-L}	A_V	Ref.
HD 45677	1.2	170	6.0	0.9	1
	0.9	30			6
HD 50138	1.0	159	3.8	0.5	1
	0.9	159			4
3 Pup	1.7	90	2.6	0.4	1
	1.53	92			4
HD 87643	0.6	134	4.8	2.8	1
	0.6	135			6
	0.8–1.2	155			4
CPD-52°9243	5.8	34	3.0	5.6	1
HD 326823	2.2	33	2.1	3.4	1
MWC 300	4.8	59	5.4	4.0	2
MWC 342	1.0	104	4.0	4.5	2
	0.2–3.8	-			7
MWC 349	8.0	166	3.7	9.3	2
	3.7-11.0	165			8
MWC 623	2.7	54	3.0	4.4	2
MWC 645	6.1	168	5.6	3.9	2
MWC 922	5.7	79	5.7	6.6	2
MWC 939	2.5	123	6.4	2.2	2
MWC 1055	3.2	51			2
HD 316285	3.3	157	1.0	6.3	3
	3.15	166			4
CD-24°5721	0.8	28	3.6	2.2	4
CPD-57°2874	2.0–3.2	47	2.1	6.2	4
MWC 930	3.9	57	0.3	9.0	5
MWC 84	2.6–5.2	160	5.3	3.7	5

A_V is the overall extinction estimated from optical color–indices. p_V and Θ_V are the polarization and position angle in the V-band. References: 1 – Barbier & Swings (1982), 2 – Zickgraf & Schulte–Ladbeck (1989), 3 – Hutchingson et al. (1994), 4 – Yudin & Evans (1997), 5 – Bergner, Miroshnichenko, Yudin (unpublished), 6 – Gnedin et al. (1992), 7 – Bergner et al. (1990), 8 – Yudin (1996). In Refs. 5 and 8 the data in the R-band are given.

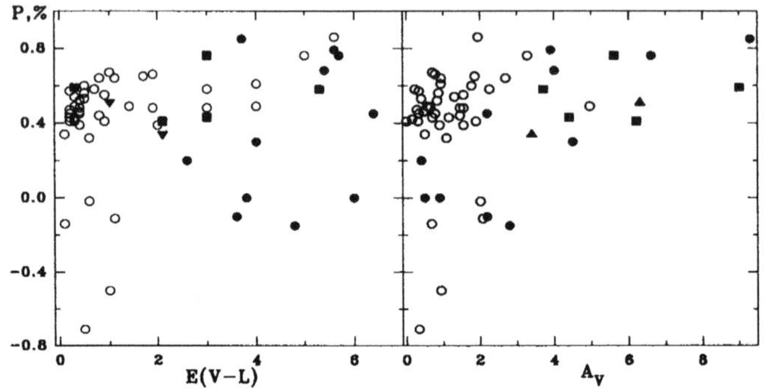

Figure 7. Polarization of B[e] stars (possible binaries – filled squares, supergiants – filled down triangles, others – filled circles). HAEBEs shown for comparison are denoted by open circles.

Long–term polarimetric observations of FS CMa in 70s were carried out by Coyne & Vrba (1976). They found polarization to be variable with time and increasing towards longer wavelengths. All observed phenomena were interpreted as supporting the model of a dust ring surrounding several gaseous regions in the extended circumstellar disc around the star. Spectropolarimetry in the UV obtained with the WUPPE during the *Astro-1* mission in 1990 showed a near–UV position angle flip by 90° and a steep rise of polarization toward shorter wavelengths (Schulte–Ladbeck et al. 1992). This provides new evidence for a disc–like geometry of the circumstellar dust around FS CMa.

Another star has been observed by WUPPE during the *Astro-2* mission in 1995 is HD 50138. It shows an almost flat polarization spectrum with a raise toward shorter wavelengths and without any position angle flip (Bjorkman et al. 1996). The UV polarization indicates the presence of the intrinsic component due to a non-spherical envelope while the absence of the flip can be due to a less dense circumstellar disk than those of FS CMa.

Zickgraf & Schulte–Ladbeck (1989) studying wavelength dependences of polarization of 8 faint B[e] stars concluded that data for two of them, MWC 342 and MWC 623, can be explained by Mie scattering, other two, MWC 645 and MWC 939, by Thomson scattering, while 3 stars: MWC 300, MWC 922, and MWC 1055, probably have no intrinsic polarization. It is followed from the above consideration that MWC 300 and MWC 922 show significant IR–excesses which can not exclude the presence of circumstellar polarization in these objects. Ivezić et al. (this volume) found that all reddening of MWC 922 is circumstellar. This partially explains why polarization of stars in its vicinity (according to Zickgraf & Schulte–Ladbeck) is rather small (~ 1.5 %).

For the last star in the Zickgraf & Schulte–Ladbeck sample, MWC 349, some intrinsic polarization has been suspected. Recently Yudin (1996) studied photometric and polarimetric behavior of this object using quasi simultaneous observations obtained in 1985 – 1992 and found strong and rapid polarization variability. It varies between 4 and 11 % in the R–band. He estimated interstellar polarization in the object's vicinity and concluded that a significant part of the observed polarization has circumstellar origin ($p_{intrinsic}$ ~ 11 % in the blue and visual regions). Yudin considers three mechanisms to be responsible for the polarization raise: Mie scattering in a dusty disk, Thomson scattering in high–density ionized clumps located inside the disk, and interstellar polarization. Spectropolarimetry of MWC 349 in the mid–IR (8–13 μm) was obtained by Aitken et al. (1990). They found that polarization peaks near 11 μm at 1.5 % and position angle also varied with wavelength. An upper limit for the interstellar polarization at this wavelength was found to be 0.5 % which is significantly lower than the observed level. The authors concluded that grains producing the polarization appear to lie in a dense rotating circumstellar disc. This is in agreement with radio data and speckle interferometric results.

Highly variable polarization has been detected earlier for MWC 342 (Bergner et al. 1990). Its changes were found to be quasi–regular with a period of 66^d (after subtraction the interstellar component) which is half a period of photometric changes in the U–band (Fig.8). Such a behavior is predicted for binary systems containing a compact object and a normal star (Dolginov et al. 1979). This prompted Miroshnichenko (1991) to suggest that MWC 342 was a Be/X–ray system.

Thus, the polarimetric results can be summarized as follows:

1. All B[e] stars observed polarimetrically display a significant degree of polarization. In many cases it can not be explained by the interstellar mechanism only.
2. Mie scattering, as well as Thomson scattering, can account for the observed wavelength dependence of polarization, so that neither gaseous nor dusty component can be ignored in analysis of observations.
3. Polarization of the objects observed many times is found to be variable. This and spectropolarimetry in some cases confirms a hypothesis of non-spherical dust distribution around B[e] stars.

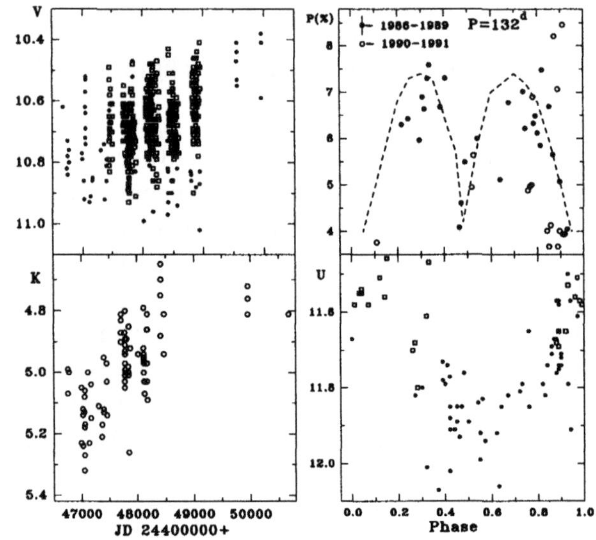

Figure 8. Light curves and polarization variability of MWC 342. Observations denoted by open squares are obtained by Bergner et al.(1995) and Kuratov et al.(this volume), those denoted by filled squares obtained by Shevchenko et al.(1993)

4. INTERFEROMETRY

Two galactic B[e] star from the sub-sample considered here have been observed by speckle interferometry: MWC 349 (Leinert 1986) and HD 45677 (Dewarf & Dyck 1993). MWC 349 was resolved at the K- and L-bands in the east–west direction with a size of 85±19 milliarcsec and barely resolved north-south at 38±18 milliarcsec. This was interpreted as a flattened circumstellar dust distribution viewed edge–on and is roughly perpendicular to the bipolar axis seen in radio. The observed optical and IR emission indicates that only 3–6 % of stellar luminosity is intercepted by dust suggesting that the obscuring disc is relatively thin in comparison with the radius of a central cavity.

HD 45677 has been resolved at a wavelength of 4.8 μm with FWHM of $0.1 \pm 0.02''$. Using distance measured by HIPPARCOS (350 ± 50 pc, ESA 1997) one can estimate linear dimension of the source of 35 A.U. Dewarf & Dyck introduced a diagram displaying an IRAS color–index [12]–[25] versus circumstellar cloud diameter in which sources with redder color–indices have in general more extended envelopes. The authors' interpretation that sources with larger size circumstellar halos are in earlier stages of evolution while sources with bluer colors are in later stages is in agreement with predictions of evolutionary models by Adams et al. (1987).

5. CONCLUSIONS

Study of photometric, polarimetric, and speckle interferometric results obtained for a group of galactic B[e] stars not identified as objects at well-known evolutionary stages (nearly 50% of initial sample) shows that rather small number of them have been investigated in detail. At the same time, even the well-observed objects are not clearly understood because of their complex structure. At the moment, we can make the following conclusions regarding all objects which can be classified as B[e] stars.

1. The majority of galactic B[e] stars locate outside of star forming regions and, hence, more likely are evolved stars rather than pre-main-sequence ones. Lack of their far-IR radiation with respect to that of near-IR could be explained as a sign of recent dust formation in their envelopes as well as of the absence of distant remnants of protostellar phase usually seen around young stars.
2. Basically, mechanisms of dust envelope formation suggested by the discoverers have been confirmed by follow up studies. Observations obtained over the last 20 years made it possible to study these processes in more detail, however additional efforts are required to determine physical parameters of the objects using up-to-date models.
3. There is a lack of high-quality observations of galactic B[e] stars. Variability of their various characteristics requires coordinated long-term observations using different observational techniques. It is extremely important to obtain long-term mid- and far-IR observations which could tell us more about circumstellar envelope evolution.
4. A large reservoir of new B[e] star candidates is a list first proposed by Dong & Hu and examined in a catalogue by Thé et al. (1994). Optical and near-IR photometry and at least medium-resolution spectroscopy is needed to preliminary classify these objects.

Acknowledgements

The author is very grateful to E. Halbedel, D. de Winter, R. Yudin, V.P. Arkhipova for permission to use their data prior the publication as well as to T. Lloyd Evans, D. Kilkenny, F. van Wyk, and F. Marang for obtaining photometric observations of some B[e] stars at the SAAO. I also thank K.S. Kuratov, D.B.Mukanov, T.A.Shejkina, R.R. Nazyrova, and B.Kh. Kallemov for assistance during observations at the Assy and Tien-Shan Observatories in Kazakhstan.

References

Adams, F.C., Lada, C.J., Shu, F.H. (1987), *ApJ*, **312**, 788

Aitken, D.K., Smith, C.H., Roche, P.F., & Wright, C.M. (1990), *MNRAS*, **247**, 466

Allen, D.A., Swings, J.P. (1976), *A&A*, **47**, 293

Arkhipova, V.P. (1975), *Variable stars*, **18**, 289

Arkhipova, V.P., Ipatov, A.P. (1982), *Sov.Astr.Let*, **8**, 554

Arkhipova, V.P., Ikonnikova, N.P., Esipov, V.F., Noskova, R.I. (1996), *Pis'ma v Astron. Zhurn*, **22**, 526

Barbier, R., Swings, J.P. (1982), Proc. IAU Symp. **98**, 241

Bergner, Yu.K., Miroshnichenko, A.S., Sudnik, I.S., et al. (1990), *Astrofizika*, **32**, 203

Bergner, Yu.K., Miroshnichenko, A.S., Yudin, R.V., et al. (1995), *A&AS*, **112**, 221

Bjorkman, K.S. et al. (1996), *ASP Conf.Ser.*, **97**, 197

Carlson, E., Henize, K.G. (1979), *Vistas in Astronomy*, **23** , 213

Coyne, G. & Vrba, F.J. (1976), *ApJ*, **207**, 790

Dewarf, L.E. & Dyck, H.M (1993), *AJ*, **105**, 2211

de Winter, D., Thé, P.S., & Pérez, M.R. (1994), *ASP Conf.Ser.*, **62**, 413

de Winter, D., van den Ancker, M.E. (1997), *A&AS*, **121**, 275

Dolginov, A.Z., Gnedin, Yu.N., Silant'ev, N.A. (1979), *Propagation and polarization of radiation in space.* Nauka, Moscow

Dong, Y.S., Hu, J.Y. (1991), *Chin. A&A*, **15**, 275

Drew, J.E. et al. (1997), *MNRAS*, **286**, 538

ESA (1997), *The Hipparcos and Tycho Catalogues*, ESA SP-1200

Finkenzeller, U. & Mundt, R. (1984), *A&AS*, **55**, 109

Gottlieb, E.W., Liller, W. (1978), *ApJ*, **225**, 488

Gnedin, Yu.N. et al. (1992), *Sov. Astr. Let.* **18**, 454

Grady, C.A., et al. (1993), *ApJ*, **415**, L39

Grady, C.A., et al. (1996), *A&AS*, **120**, 157

Halbedel, E.M. (1989), *PASP*, **101**, 999

Hutchingson, M.G., Albinson, J.S., Barrett, P., et al. (1994), *A&A*, **285**, 883

Kazarovets, E.V. & Samus, N.N. (1997), *IBVS* **4471**

Kilkenny, D., Whittet, D.C.B., Davies, J.K., et al. (1985), *SAAO Circ.*, **9**, 55

Leinert, C. (1986), *A&A*, **155**, L6

Manfroid, J., Sterken, C., Bruch, A., et al. (1991), *A&AS*, **87**, 481

McGregor, P.G., Hyland, A.R., Hillier, J. (1988), *ApJ*, **324**, 1071

Miroshnichenko, A.S. (1991), *Infrared spectral region of stars.* Proc. Int. colloq. held in Montpellier. Jaschek, C., Andrillat, Y. (eds.), p. 163

Miroshnichenko, A.S. (1994), *ASP Conf.Ser.*, **62**, 134

Miroshnichenko, A.S. (1995), *A&ATr*, **6**, 251

Miroshnichenko, A.S., Kuratov, K.S., Ivezić, Ž., & Elitzur, M. (1997), *IBVS* **4506**

Parthasarathy, M. & Pottasch, S.R. (1989), *A&A*, **225**, 521

Pogodin, M.A. (1997) *A&A*, **317**, 185

Schulte–Ladbeck, R.E. et al. (1992), *ApJ*, **401**, L105

Sitko, M., Halbedel, E.M., Lawrence, G.F., et al. (1994), *ApJ*, **432**, 753

Sterken, C., Manfroid, J., Anton, K., et al. (1993), *A&AS*, **102**, 79

Sterken, C., Manfroid, J., Beele, D., et al. (1995), *A&AS*, **113**, 31

Thé, P.S., de Winter, D., & Pérez, M.R. (1994), *A&AS*, **104**, 315

Torres, C.A.O. et al. (1995), *AJ* **109**, 2146

van der Veen, W.E.C.J. & Habing, H.J. (1988), *A&A* **194**, 125

Wackerling, L.M. (1970), *Mem.Roy.Astron.Soc.*, **73**, 153

Wolf, B. & Stahl, O. (1985), *A&A*, **148**, 412

Yudin, R.V. (1988), *Sov.Astron.* **32**, 652

Yudin, R.V. (1996), *A&A* **312**, 234

Yudin, R.V. & Evans, A. (1997), *A&AS*, (submitted)

Zickgraf, F.-J., Wlof, B., Leitherer, C., Appenzeller, I., Stahl, O. (1986), *A&A*, **163**, 119

Zickgraf, F.-J. & Schulte–Ladbeck, R.E. (1989), *A&A*, **214**, 274
Zickgraf, F.-J. & Stahl, O. (1989), *A&A*, **223**, 165

Discussion

Cassinelli: The radial dependence of the density, $\rho \propto r^{-n}$ is very shallow ($n \leq 1$). Could this be an indication that the outer dust shell is disconnected, as for example might occur if there were segmented mass loss events?
Miroshnichenko: It depends on the evolutionary state of a star. In pre-main-sequence objects this could mean that we see distant dust from the protostellar cloud, while in post-main-sequence stars recurrent mass loss events cannot be ruled out.
Bjorkman: Do you see any systematic differences in the IR excesses of the pre-main-sequence candidate B[e] stars (such as HD 45677) and the supergiant B[e] stars?
Miroshnichenko: It is not easy to answer this question because in some cases we are not sure in an object luminosity estimate. Many of the known B[e] dwarfs and supergiants display both near- and far-IR excess, but sometimes supergiants can show the far-IR excess only (for example MWC 349).
Waters: You compare LS II+22°8 to 3 Puppis. The latter star is a spectroscopic binary with a prominent circumbinary dust shell (disk), with T_{dust}=800 K. Does your object show similar properties?
Miroshnichenko: I used 3 Pup only for comparison of Hα profiles. LS II+22°8 is not known as a binary and, as I have shown, probably has no dusty shell.
Bjorkman: One of the striking differences between Magellanic and galactic B[e] stars is their level of polarization. Do you think that most of the large polarization level in galactic B[e]'s is due to interstellar polarization, or are there systematic differences between the two groups?
Miroshnichenko: We do see a correlation between polarization of B[e] stars and overall, inter- and circumstellar extinction toward them. No conclusion can be drawn for a relation between the polarization and any of these components before we disentangle them. However, it can be seen from our poster (Ivezic et al, this meeting) that for instance all reddening for MWC 922 can be explained by a circumstellar component only. At the same time Zickgraf and Schulte-Ladbeck(1989) have detected very high polarization for this star with almost an interstellar wavelength dependence.
Houziaux: Could you mention if the type of photometric variability found in B[e] galactic supergiants is similar to the variability observed for B[e] stars in the Magellanic Clouds? or are these variabilities of different type?
Miroshnichenko: Objects suggested to be B[e] supergiants in the Galaxy

show different photometric behavior. HD 316285 shows almost no changes for many years, something which is similar to the behavior of the Magellanic Cloud B[e] supergiants. The other star, MWC 349, exhibits cyclic variability. Perhaps they are at different evolutionary stages.

Hubert: Is there any detection of rapid variability possibly linked with the rotational period or phenomena such as non-radial pulsations?

Miroshnichenko: Some rapid changes of brightness were detected by D.de Winter for Hen 1191 (ASP Conf.Ser. N° 62 ,1994, p.413). Rapid variability was also detected in HD 45677 but its amplitude was very small, several tenths of a magnitude.

Marlborough: You suggested the galactic objects were more variable than those in the Magellanic Clouds. Is there some physical reason for this difference?

Miroshnichenko: Galactic B[e] stars are a mixture of objects of different evolutionary stages. So we can expect very different amplitudes and types of variability for them. My result is just an average for the galactic sample which is still small.

Zickgraf: You said that the average photometric variability of the B[e] stars has an amplitude of about 0.5 magnitudes. Can you give the average variability for the different subclasses? Are there differences for the subclasses?

Miroshnichenko: I can not split the well observed objects into subsamples because of small number statistics. At the moment we can say that HD 45677 and HD 50138 show a similar type of variability. Other stars differ among them.

Lamers: If the variability is due to orbiting circumstellar clouds you expect a clear change in colors associated with it: the fainter, the redder. Do you observe this?

Miroshnichenko: The variability of B[e] stars is different from the one of the Herbig Ae/Be stars. The latter show Algol type minima lasting for several days. B[e] stars exhibit slower changes. Moreover their color changes are very small. The only star showing reddening with brightness decrease is HD 87643, but it decreases for a very long time.

OBSERVATIONS OF GALACTIC B[e] STARS AT THE TIEN–SHAN OBSERVATORY

KURATOV KENESKEN S.
Fesenkov Astrophysical Institute of the Kazakhstan National Academy of Sciences, Almaty, 480068, Kazakhstan

AND

MIROSHNICHENKO ANATOLY S. AND BERGNER YURI K.
Central Astronomical Observatory of the Russian Academy of Sciences at Pulkovo, Saint-Petersburg, 196140, Russia

B[e] stars is a group of emission–line early–type stars with large infrared excesses which can be due to circumstellar dust radiation. It contains objects of different physical nature and requires comprehensive observations to understand they intrinsic properties. Very important information about the latters comes from multicolour photometry, however many of the galactic B[e] stars have been rarely observed in the past. We have been performing a photometric study of the northern B[e] stars at a 1–meter Zeiss reflector of the Tien–Shan Observatory in Kazakhstan since 1995. This is a continuation of our program carried out in 1989–1994 which resulted in more than 200 *UBVRIJHK* observations of 10 objects obtained at the Assy Observatory in Kazakhstan (Bergner et al. 1995).

Our new program includes classical B[e] stars from the list of Allen & Swings (1976) - HD 45677, HD 50138, MWC 342, MWC 349, MWC 922 - and possible candidates to this group distinguished by their emission–line spectrum and excesses at the IRAS wavelengths - AS 78, AS 116, AS 117, AS 314, AS 319, AS 321, LS II +22° 8, MWC 657, MWC 953 (Dong & Hu 1991). Nearly 100 Johnson *UBVRIJHK* observations of these objects have been obtained so far with a two–channel photometer–polarimeter (Bergner et al. 1988) using a 26″ diaphragm. Here we report first results of these observations. Optical data for MWC 657, MWC 922, AS 319, AS 321 as well as near–IR data for MWC 657, MWC 953, AS 314 were obtained for the first time. Results for two objects, LS II +22° 8 and AS 78, are discussed by Shejkina & Miroshnichenko (this volume).

A. M. Hubert and C. Jaschek (eds.), B[e] Stars, 165–166.

The largest IR–excesses in our sample have FS CMa and MWC 922. Their SEDs are similar except for the type of a 10 μm silicate feature. It is in emission in FS CMa which implies an optically thin circumstellar dusty envelope while in absorption in MWC 922 indicating optically thick dust. Since the overall extinction at the line of sight towards MWC 922 is not high one can suggest that its circumstellar envelope is non–spherical.

Four objects – FS CMa, HD 50138, MWC 657, and MWC 342 – display very similar SEDs with a fast decrease toward longer wavelengths. They have almost the same density law in their envelopes but different optical depths (see Ivezić et al. in this volume). The former two stars have luminosities close to the ZAMS and could be young or evolving to the planetary nebula phase. The same state could also have MWC 657 (Miroshnichenko et al. 1997). MWC 342 which is brightening in the visual and near–IR needs in further photometric monitoring as well as in high–resolution spectroscopy to draw more certain conclusions on its evolutionary state.

Two objects, MWC 349 and AS 314, are thought to be supergiants. The former is an O– or an early–B star with optically thin silicate deficient dusty envelope. The He II 4686 Å line sometimes appears in emission in its spectrum. A similar spectroscopic behaviour shows a B[e] supergiant Hen S18 which was suggested to have a companion (Zickgraf et al. 1989). AS 314 displays a small near–IR excess probably due to free–free emission and a large far–IR one which can be explained by the presence of cool dust in its outer envelope. Such dust could originate during outbursts in the past like those in the LBVs (WRA 751, HR Car).

Other stars – AS 116, AS 117, AS 319, AS 321, and MWC 953 – are almost unstudied. The IR–excesses of the former two objects are similar to those of Herbig Ae/Be stars. AS 116 retains a large near–IR excess which is observed only in several such objects with early–B spectral types (e.g., MWC 137) and, hence, could be very young. AS 117 looks more evolved because of the absence of the near–IR excess (Miroshnichenko et al. 1996). The latter three stars display only far–IR excesses and it is difficult to judge about their nature without spectroscopic data.

References

Allen, D.A., Swings, J.P. (1976) *A&A*, **47**, 293

Bergner, Yu.K., Bondarenko, S.L., Miroshnichenko, A.S. et al. (1988) *Izvestia Glavn. Astron. Obs. v Pulkove*, **205**, 142

Bergner, Yu.K., Miroshnichenko, A.S., Yudin, R.V., et al. (1995), *A&AS*, **112**, 221

Dong, Y.S., Hu, J.Y. (1991) *Chin. A&A*, **15**, 275

Miroshnichenko, A.S., Bergner, Yu.K., Kuratov, K.S., Mukanov, D.B., Shejkina, T.A. (1996) *Astronomy Reports*, **40**, 509

Miroshnichenko, A.S., Kuratov, K.S., Ivezić, Ž., & Elitzur, M. (1997) *IBVS*, **4506**

Zickgraf, F.-J., Wolf, B., Stahl, O., & Humphreys, R.M. (1989) *A&A*, **220**, 206

POLARIMETRIC VARIABILITY OF SOME PECULIAR Ae/Be STARS

N.G. BESKROVNAYA
Pulkovo Observatory, 196140 Pulkovo, St.Petersburg, Russia

The envelopes of young Herbig Ae/Be stars are known to contain local inhomogeneities in their gaseous envelopes (Beskrovnaya et al. 1995). The study presented in this poster was aimed at a search for azimuthal inhomogeneities in the envelopes of two objects with uncertain classification by means of polarimetric method. In more detail the results of this study, uncluding high-resolution spectroscopy, can be found in Beskrovnaya et al. (1997a,b).

The emission line star HD 163296 was first classified as a peculiar early-type star (Allen & Swings 1976) and later on as a young Herbig Ae/Be star (Finkenzeller & Mundt 1984). *VRI* polarimetric observations of HD 163296, obtained in July 17-31, 1995 at the SAAO, give evidence of the presence of at least two variable linear polarization components.

We have found the nightly mean values of the u-parameter to be satisfactorily approximated by a sinusoidal curve with a period $P = 7.5 \pm 0.3$ days, superimposed on a linear trend with the same slope in all three bands and the mean value increasing towards longer wavelengths. Variations of the q-parameter are of essentially lower amplitude. After subtraction of the trend, the variations of the q parameter in the VR-bands can be fitted by a sinusoidal curve with the same period (7.5 days) and $\pi/4$ phase shift relative to the corresponding u-curves, while in the I-band the sinusoidal variations disappear.

Sinusoidal component of variability of the q, u Stokes parameters in V and R passbands can be caused by the orbital motion of a dust or gas condensation scattering non-selectively the stellar light. However, the model involving a dust condensation cannot explain the non-sinusoidal character of the Q parameter variations in the I passband. That is why we suggest to interpret the observed phenomenon assuming that a rotating inhomogeneity arises in the gaseous media in the presence of magnetic field.

A. M. Hubert and C. Jaschek (eds.), B[e] Stars, 167–168.

The results of calculations (for details see Beskrovnaya & Pogodin 1998) for the following model parameters: the orbital period $P = 15^d$, $i = 78°$, and $B = 120\,$G with the lines of force along the radius-vector, are in good agreement with the observed polarimetric curves and demonstrate that this model can explain phase shifts on the q, u-curves as well as the disappearance of the second harmonic of the Q-parameter variations in the I-band, where the influence of magnetic field is stronger.

The Ae-star HD 36112 remained one of the poorly–studied bright emission–line objects untill recent time. The *UBVRI* polarimetric observations of HD 36112 had been collected in 1994–1996 at the CrAO. The normalized Stokes parameters demonstrate a striking variability from date to date. On the (q, u)-plane the nightly mean values for each passband form a cloud of points with the radius of scattering upto 0.4% that exceeds the errors of individual points by an order of magnitude.

The positions of these clouds on the (q, u)-plane exhibit a conspicuous wavelength dependence. The points corresponding to the weighted mean values for all the data in each passband are situated along a straight line. This allows to distinguish by means of least-squares method two constant background components of polarization: a non-selective component with $P = 0.18\%$ and $\theta = 156° \pm 3°$, and a selective one with $P(\lambda) \propto \lambda^{-1}$ and $\theta = 59\overset{\circ}{.}4 \pm 0\overset{\circ}{.}6$.

Variations of the q and u parameters in the B-passband during four consecutive nights revealed the presence of the long-term trend, which has a form of the arc on the (q, u)-plane. The residual variations after subtraction of the trend exhibit similar periods in both parameters ($P = 0\overset{d}{.}9 \pm 0\overset{d}{.}02$ for the q-parameter, and $P = 0\overset{d}{.}86 \pm 0\overset{d}{.}01$ for the u-parameter). The intricate picture of polarimetric variations during four nights can be connected with rotation of a jet-like inhomogeneity with $P = 42^h$, and orbital movement of a more remote condensation.

Acknowledgements

This work was sponsored by the INTAS under grant 93-2478-ext.

References

Allen D.A. and Swings J.P. (1976), *A&A*, **vol. no. 47**, p. 293

Beskrovnaya N.G., Pogodin M.A., Najdenov I.D., Romanyuk I.I. (1995), *A&A*, **vol. no 298**, p. 585

Beskrovnaya N.G., Pogodin M.A., Yudin R.V., Franco G., Vieira S., Evans A. (1997), *A&AS*, in press

Beskrovnaya N.G., Pogodin M.A., Miroshnichenko A.S. et al. (1997), *A&A*, to be submitted

Beskrovnaya N.G. and Pogodin M.A. (1998), *Astrofizika*, to be submitted

Finkenzeller U. and Mundt R. (1984), *A&AS*, **vol. no. 51**, p. 109

POLARIMETRY OF SOUTHERN PECULIAR EARLY–TYPE STARS

R.V. YUDIN
Central Astronomical Observatory of the Russian Academy of Sciences at Pulkovo, 196140 Saint-Petersburg, Russia

AND

A. EVANS
Department of Physics, Keele University, Keele, Staffordshire, ST5 5BG, UK

1. Introduction

An important factor is the relation of HAEBE stars with other groups of peculiar early type stars, such as Be and B[e] stars, luminous blue variables (LBVs) etc. In many cases polarimetric properties of HAEBE stars and B[e] (Be) stars are similar. We have carried out a polarimetric study of peculiar early–type stars with the following aims:

1. First, to obtain new polarimetric (linear and circular) data for a large group of HAEBE stars and candidate members, as well as for some peculiar Be and B[e] stars and LBVs. The whole list of programme stars consists of 60 objects located mainly in the southern hemisphere. For 40 of the 60 objects polarimetric data have been obtained for the first time.
2. Second, to study polarimetric variability of the programme stars on different time-scales (minutes ... hours ... days).

2. Observations

A programme of optical polarimetry of selected stars was carried out during three runs in 1994 August 2–15, 1995 January 23 – February 6 and 1995 July 18 – August 31 at the South African Astronomical Observatory (SAAO). The University of Cape Town photometer-polarimeter module was used on the SAAO 0.75-m telescope, although the observations for the period 1994

A. M. Hubert and C. Jaschek (eds.), B[e] Stars, 169–171.

August 9–15 were carried out on the SAAO 1-m telescope. During the first run of observations data in the Johnson *UBV* and Cousins R_cI_c filters were obtained. In the second and third runs only *UBV* or VR_cI_c filters were used respectively. All observations were made with a 20" aperture.

3. Conclusions

Our main conclusions may be summarized as follows:

1. There was no evidence for polarimetric variability in only 8 stars in our survey (MWC 120, MWC 137, Z CMa, RCW 34, He 3–140, HD 303004, WRA 15–689 and MWC 900) but polarimetric variability for two of these stars (Z CMa and MWC 137) has been claimed in the literature; variability on a long time-scale (years) may take place in RCW 34.
2. Analysis of the data allows us to conclude that there is no significant difference in the proportion of stars showing rapid (on a time-scale of minutes) variability between the three classes of stars (HAEBE, Be and B[e]). Such variability is present in about 40% of stars in the each class for which rapid variability has been studied.
3. Variability on a time–scale of days was detected in most of the programme stars for which such an investigation was carried out. Probably the vast majority of all HAEBE stars, as well as Be stars with enough measurements, are polarimetric variables.
4. Our data, together with the polarimetric data in the literature, enables us to deduce variability on a long time-scale (year or longer) for RCW 34, MWC 272, Z CMa, V586 Ori, MWC 925 and He 3–365.
5. Consequently variability on different time-scales is present in most of the investigated stars, in all three classes, and can not be used as a defining criterion for the classifying stars in the various groups.
6. In our survey no significant circular polarization but at $\geq 3\sigma$ level was detected in MWC 120, RCW 34, He 3–554, HD 37411, LkHα 115 and He 3–365. Polarimetric observations with a good accuracy of HD 163296 showed the presence of circular polarization twice in the *V* band, three times in the R_c band and seven times in the I_c band.
7. Only four stars show a clear correlation between photometric state and polarimetry. These are V350 Ori (probably a HAEBE star with Algol–like minima), VY Mon (a HAEBE star or FU Ori type star), the peculiar star WRA 15–535 and the TTS RY Lup; for these objects the polarization was larger when the star was faint. For other stars which showed polarimetric variability the variations are not accompanied by photometric changes.

Discussion

Bjorkman: What was the level of circular polarization you detected in HD 163296 and what mechanism do you propose for producing the circular polarization.

Yudin: The weighted mean values of circular polarization for the star were detected above the 3 σ level as follows: $Pc(v) = 0.012 \pm 0.003$ %, $Pc(r) = 0.020 \pm 0.005$ %, $Pc(i) = 0.025 \pm 0.005$ %. These results imply the presence of aligned particles in dusty envelopes, which are located at larger distance from the star than that of the region where linear polarization originates. Thus the mechanism of circular polarization is a conversion of linear polarization to circular by scattering on the oriented dust particles.

SUB-MILLIARCSECOND STRUCTURES IN P CYG

F. VAKILI, P. BERIO, D. BONNEAU, O. CHESNEAU

AND

D. MOURARD, P. STEE AND N. THUREAU

Observatoire de la Côte d'Azur, UMR-CNRS 6528, 06460 Saint Vallier de Thiey, France

Abstract. We report the first resolution of the envelope of P Cyg in Hα using the GI2T spectro-interferometer in the southern France. We detected a localized asymmetry at .8 mas to the South of P Cyg's photosphere on August 1994. Such observations should be carried systematically on other hot, massive stars to ascertain the physical parameters of their strong winds. The goal is to calibrate the Wind Momentum-Luminosity Relation among WR and O,B,A supergiant stars for their future use as cosmological candles.

Key words: Stars: Massive stars – Stars: Stellar Interferometry

1. Optical Interferometry of P Cyg with the GI2T

This section summarizes the milliarcsecond (mas hereafter) optical interferometry of the LBV star P Cyg with the GI2T (Mourard et al. 1995) at the OCA. On August 4th 1994 we obtained interferometric data in Hα spectral region with an interferometric baseline of 17m7. Based upon robust calibrations on 2 reference stars we estimated the equivalent uniform disk diameter of P Cyg in Hα as 5.52 $\pm$ 0.47 mas. Furthermore, we sampled the interferometric signal across Hα within 16 narrow spectral channels, each .17nm wide, starting from the blue and finishing in the red continua. Thus we could determine the position of P Cyg Hα emission relative to its photosphere versus Doppler-shift across the line profile (for more details see Vakili et al. 97). We found that the N-S projected position of P Cyg at a Doppler-shift of -208 km/s is by 4 stellar radii to the South of the

A. M. Hubert and C. Jaschek (eds.), B[e] Stars, 173–176.

photosphere whilst at all Doppler-shifts, from the blue to the red wings of the line, the position of P Cyg is constant and centered on the underlying star (see figure 1 and 5 of reference Vakili et al. 1997). This is interpreted as the first detection and clear evidence for a localized asymmetry in the wind of this famous LBV.
It is interesting to monitor such events in the future at high spatial resolution in order to follow their propagation out to larger scales on regard of dynamical evolutionnary models (Garcia et. al. 1996). These models predict enhanced density asymmetries within a few parsecs when LBVs evolve towards the WR stage. The interferometric monitoring demands indeed multi-scale spatial resolution observations, from sub-mas to tens of arcseconds. For the latter range of spatial structures, detecting weak nebulosities around the central engine and their temporal variability implies a challenging problem of high dynamic range imaging. From the ground, this could be only tackled by coronographic adaptive optics on large monolithic telescopes accompanied by spectrally resolved decametric interferometry and simultaneous high resolution spectroscopy.

2. Towards the Calibration of the Wind Momentum-Luminosity Relation: WN-LR

The theory of radiation driven winds (Kudritzki et al. 1994 and references therein) predicts a relation between the momentum contained in the wind and the stellar luminosity which for the Milky Way (MW) metallicity is expressed as:

$$\dot{M} v_{\infty} R^{0.5} \propto L^{x} \tag{1}$$

where $\dot{M}$ is the mass-loss rate, v_{∞} the wind terminal velocity, R the photospheric radius, L the luminosity and x a metallicity parameter estimated as 1.5 for the MW. All these parameters could statistically be determined for galactic sources, LMC, SMC and for close galaxies like M31 or M33 using low resolution spectroscopy of Hydrogen, Helium or other lines depending on the spectral type and on the basis of hydrodynamic model atmospheres of hot luminous stars. Other photometric data in the UV, IR or radio wavelengths could refine the empirical parameters of the relation.
Direct measurements of geometrical parameters such as the angular value of R in equation (1) and wind parameters like v_{∞} or $\dot{M}$ constrained by long baseline interferometry in Balmer or HeI, HeII lines (Stee et al., 1995) should bring independent improvements on the WM-L Relation and its use for cosmological distance calibrations. Patrol interferometric observations on the GI2T are mandatory to pave the route of future runs on the VLTI where the situation is favored by accessing to LMC and SMC. Sub-mas observations in the visible with the VLTI 8m telescope array and its improved

sensitivity should enable us to persue the work reported here on the wind of P Cyg to a representative sample of hot luminous stars in Magellanic Clouds.

Acknowledgements

this work was partially supported by CNRS/GDR-MC and PNHRAA from INSU-CNRS.

References

Garcia-Segura G., Lac Low M-M., et al., 1996, A&A, 305, 229-244

Kudritzki R.P., Lennon D.J. and Puls J., 1994, proceedings of ESO Workshop "Science with the VLT", eds. Walsh J.R. and Danzinger I.J., Springer Verlag Heidelberg, pp 246-255.

Mourard D., Bosc I., et al., 1994, A&A, 283, 705.

Stee P., De Araujo F.X., et al., 1995, A&A, 300, 219.

Vakili F. Mourard D, et al. 1997, A&A, 323, 183-188.

Discussion

Lamers: I have seen new unpublished very high resolution radiomaps of P Cyg. They show blobs and clumps which are surprisingly bright and variable. This may be related to the blob that produced the dip in the phase map that you showed.

Vakili: I emphasize that the blob detected by the GI2T is a few stellar radii distant from P Cyg photosphere or less than 1 milliarcsecond in terms of angular separation. I wonder whether the radio map you mention is comparable in angular scale to our observations since only intercontinental VLBI achieves milliarcsec resolutions at a few GHz.

Cassinelli: Couldn't you have a situation in which there are many blobs (comets) around the star? I say this because to see the feature you did, the blob would have to be in a very restricted portion of the flow.

Vakili: Our observations do not say anything about the number and size of blobs/comets. They just say that they should have the same radial velocity and be compactly grouped to give a single dip otherwise there would produce different dips in the phase map at different Doppler-shifts.

Owocki: It is peculiar that this blob appears on the absorption side. This must imply that it is extra emission that is filling in the absorption trough, which must be formed along the line of sight to star. This blob must be displaced from that column, in front of the star. Is this a correct interpretation?

Vakili: Yes the blob has the same radial velocity as the projected absorbing column of the wind against the photosphere but is spatially at a different position as projected on the sky. Whether this corresponds to extra emission or a localized instability in the density/velocity/temperature structure of the wind needs more than a single isolated detection as the result presented here.

Zickgraf: Could the blob you see be a hot spot or an accreting companion?

F. Vakili: To be correct, your suggestion would need the hot spot or the companion be located at a few stellar radii from the photosphere. Whilst your first scenario cannot be ruled out with the present observations the second would imply, I assume, strong X-ray emission due to the close contact of the companion and P Cyg itself. One must check in the literature whether simultaneous observations have been carried in those wavelengths in August 1994.

Bjorkman: P Cygni is polarimatrically variable with a random position angle for the variation. Can you determine the position angle of the Hα perturbation? If so, it would be interesting to compare to the polarization PA to see if they agree. This would support your hypothesis of an ejected blob (that produces excess electron scattering). Second, do you have enough coverage in the (u,v) plane to determine if P Cygni is asymmetric in its Hα emission?

Vakili: The configuration of the GI2T baseline did not permit to check for such correlations since the baseline is North-South so that only projected position of asymmetries could be determined. However the situation should soon change since we are currently implementing new components in the interferometer so as to obtain 2D (u,v) coverage using Earth-rotation synthesis pending the arrival of a third West 1m5 telescope to perform true aperture synthesis.

MODELS FOR THE WINDS AND DISKS AROUND B[e] STARS

J. P. CASSINELLI
Astronomy Department, University of Wisconsin
Madison, WI, 53706

1. Introduction

The basic picture for the disks around B[e] stars was first developed by Zickgraf et al (1985). This is a review of several of the models that have been proposed to explain the disks around rapidly rotating luminous B stars. Circumstellar disks are also present around forming stars, however those are "inflowing" or accretion disks. For the B[e] stars, we are interested in "outflowing disks" associated with the mass loss from the stars.

The outflowing disk models can be separated into two classes. First, there are 1 D models for which an enhanced density is produced in the equatorial sector of the star. For the 1 D models the flow is assumed to be primarily radial and no flow occurs from one latitude zone to another. Second, there are the 2 D models for which the enhancement in the density in the equatorial region arises primarily from the flow of material down from higher latitude zones. Magnetic fields can play a role in both types of models. The possibility for measuring the fields from observations of UV wind lines of the stars is discussed briefly. Table 1 summarizes the various models that have been considered recently for producing outflowing disks. Here we will consider the luminous magnetic rotator and the bi-stability models as examples of 1 D models for B[e] disks. We will discuss wind compression and the possibility of magnetic channelling as examples of 2 D models.

2. The One Dimensional Sector Models

The basic problem addressed by 1 D models is how rotation can lead to a factor of 10 enhancement in the density of the radial streams of equatorial outflow. The B[e] stars are luminous, and hence they would produce a wind with a rather large mass loss rate even if they were not rotating. This

A. M. Hubert and C. Jaschek (eds.), B[e] Stars, 177–188.

TABLE 1. Outflowing Disk Models

Models	Comments	References
Sectored 1 D models:		
Luminous Magnetic Rotators	$B_{r,o} \approx 100$ G	Poe et al. 1989
Pulsation + Viscosity	Keplerian disks	Lee, Saio, & Osaki, 1991
Bi-Stability Model	gravity darkening effect	Lamers & Pauldrach, 1991
Equatorward Flow 2 D Models:		
Wind Compressed Disks	shock compressed disk	Bjorkman & Cassinelli, 1993
Wind Compressed Zones	non-shocked compression	Ignace et al. 1996
Magnetic Channelling	solar analogue	Charbonneau & Hundhausen, 1

means that there should be an outflow from the polar regions owing to radiation forces operating in these regions. However, because of centrifugal and rotational distortion effects, there can be an enhanced outflow in the equatorial region.

2.1. MAGNETIC ROTATOR MODELS

The Luminous Magnetic Rotator models for B[e] stars are a hybrid of the rotating solar wind model that was developed by Weber and Davis (1967) plus the line driven wind models of Castor, Abbott, and Klein (1975) (CAK). This combination of driving forces was first considered for rapidly rotating O stars by Friend and MacGregor (1984). It was later generalized to account for the finite disk correction of modified CAK (or mCAK) theory by Poe and Friend (1986), and they applied the model to classical Be stars. Poe, Friend and Cassinelli (1989) considered the model as a possible explanation of the Wolf Rayet "momentum problem" ($\dot{M} v_\infty >> L/c$). It was shown that a sector model could explain both the observed high v_∞ and the high $\dot{M}$ inferred from radio flux observations. The edge value on the P-Cygni profiles, that determines v_∞, arises from material in the polar sector; the high radio flux originates primarily in the denser equatorial sector. Cassinelli, Schulte-Ladbeck, Poe and Abbott (1990) used the same sectored model for the specific case of B[e] stars. Their results are shown in the $\dot{M}$ versus v_∞ plot of figure 1.

On analyzing such an $\dot{M}$ versus v_∞ plot, we find that the luminous magnetic rotator models are composed of several isolatable wind driving mechanisms. The B[e] star's basic parameters of mass, luminosity, and ef-

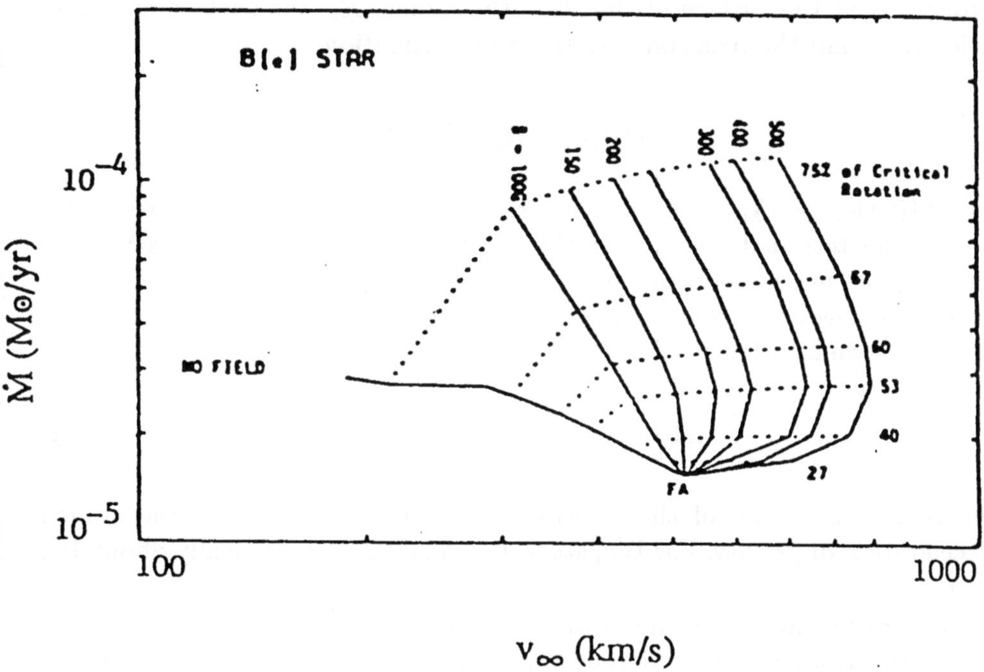

Figure 1. The mass loss rate and terminal velocities achieved by the Luminous Magnetic Rotator models of B[e] stars for a range of magnetic fields, $B_{r,o}$ and rotation rates ω. There is a basal mass loss rate, marked as FA, that is set by the line driven wind mechanism. If the magnetic field exceeds a minimal value, then the rapid rotation can increase both the terminal speed and the mass loss rate by significant amounts. (From Cassinelli et al. 1990)

fective temperature are sufficient to determining the mass loss rate and terminal velocity of of the non-rotating star using mCAK theory. In Figure 1 is shown the mCAK result of Friend and Abbott (1986). Kudritzki et al (1990) give a convenient cooking recipe for finding $\dot{M}$ and v_∞. Now lets add rotation and magnetic field to the model. If the field is smaller than a value, B_{min}, determined below, then the addition of the field has relatively little effect on the radial wind flow speed, and the wind is said to be a weak or *slow magnetic rotator* (SMR). For such rotators, the location of the star on the $\dot{M}$ versus v_∞ plot does not move from the mCAK point. Belcher and MacGregor (1976) developed *fast magnetic rotator* (FMR) theory, in which they found that if the field is above B_{min} and if the star is rotating rapidly the terminal speed of the flow will increase but at roughly constant $\dot{M}$. The increased velocity is of the FMR models is conveniently expressed in terms of the Michel velocity, V_M, which is determined by the energy deposited in the wind by the magnetic rotator forces. The Michel velocity is related to properties at the surface of the star (the radius, R_o, the surface magnetic field, $B_{r,o}$, and the rotation rate Ω), by the equation

$$\dot{M}{V_M}^3 = {R_o}^4\Omega^2{B_{r,o}}^2 \tag{1}$$

In FMR theory V_M is the fast critical point velocity, and it is also very nearly the terminal velocity of the wind. If we use in this equation the value of the mass loss rate and v_∞ derived from the mCAK theory for $\dot{M}$ and V_M respectively, and if we set Ω equal to the maximal or breakup rate, Ω_{bu}, then we can derive the value for B_{min}.

$$B^2_{min} = \dot{M}v^3_\infty/R_oGM_*(1-\Gamma) \tag{2}$$

where Γ is the ratio of the outward radiative acceleration to the inward acceleration of gravity. For B[e] stars, the value B_{min} is typically about 100 gauss.

Figure 1 shows the values of $\dot{M}$ and v_∞ that can be achieved by luminous magnetic rotators with different values of the surface field, $B_{r,o}$, and the rotation rate expressed as the ratio $\omega = \Omega/\Omega_{bu}$. In the FMR regime the mass loss rate remains at the value determined by the line driving of mCAK theory. At very large rotation rates the mass loss rate increases as the star enters the *centrifugal magnetic rotator* (CMR) regime described by Hartmann and MacGregor (1982), and by Cassinelli (1990). In the CMR regime, the mass loss rate is determined only by the rotation ratio ω, nearly independently of the field strength $B_{r,o}$ on the star. This is because the field is sufficiently strong to cause the subsonic region to rotate as a solid body and the mass loss rate is determined by the density at the critical point. Increasing the field does not change the solid body rotation of the subsonic zone, but only the distance out into the supersonic region where centrifugal effects continue to be important.

We see from figure 1 that there is a wide range of $\dot{M}$ and v_∞ values that can be produced in luminous magnetic rotator theory. Given the freedom to adjust $B_{r,o}$ and ω, the luminous magnetic rotator model certainly allows for solutions which have the density in the equatorial wind a factor of 10 larger than that in the polar (mCAK) regions. However, we need some information about the actual field strength at the surface of the star. We will return to this later. Cassinelli (1991) discusses the $\dot{M}$ versus v_∞ plot for a B[e] star which accounts for the theoretical upper and lower limits to surface magnetic fields as derived by Maheswaran and Cassinelli (1988, 1992).

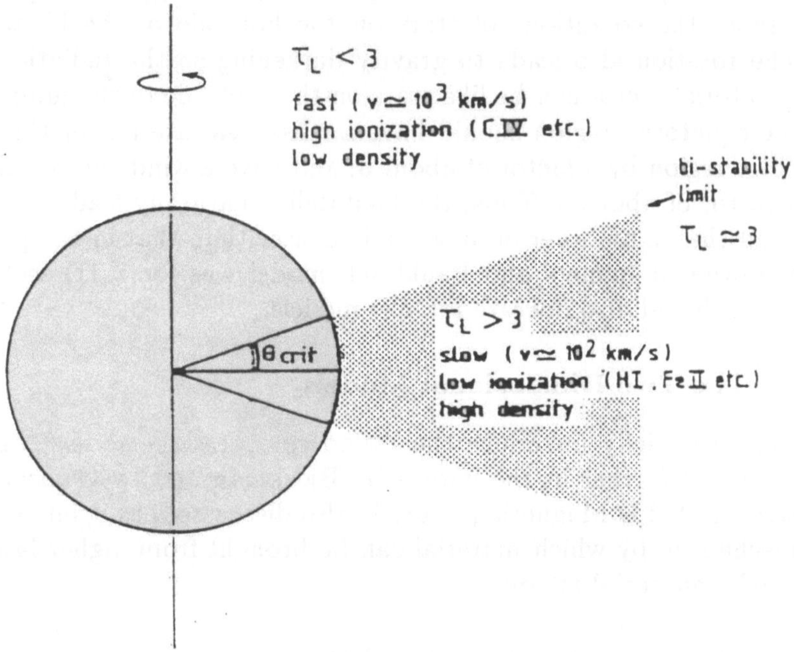

Figure 2. Illustrates the change in wind conditions from the polar sector to the equatorial sector in the Lamers and Pauldrach bi-stability model. The polar winds are faster and of lower density and tend to be driven by lines of high ionization state. In the equatorial region the EUV ionizing radiation is not present to produce the high ion stages and thus a slower denser wind is driven by lower ionization species such as Fe II. (From Lamers and Pauldrach 1991)

2.2. THE BI-STABILITY MECHANISM

Pauldrach and Puls (1990) found that for there occurs a jump in the wind model properties of B supergiants with effective temperatures near 20,000 K. For stars hotter than about this value, the winds tend to be fast and have relatively low densities. For stars cooler than this, the winds tend to be slower and denser. The reason for the difference is that winds of hotter stars are driven by EUV lines, while for stars on the cool side of the jump the wind is optically thick in the EUV, so the wind driving is produced by lines in the UV spectral region. The switch is accompanied by a change in the CAK line force parameters. and hence also by changes in the mass loss rate and terminal velocity. Lamers and Pauldrach (1991) suggested that the same effect is important from one latitude sector to another over rapidly rotating B stars. Their model is illustrated in Figure 2.

The rapid rotation leads to polar brightening, so the polar radiation mimics the conditions of stars on the hot side of the bi-stability jump. The rotation also leads to gravity darkening so the radiation field in the equatorial sector can be like stars on the cool side of the jump. As a result, the equatorial region should have a mass loss rate higher than that in the polar region by a factor of about 3, and have a wind speed that is lower by a factor of about 3. Thus, the bi-stability naturally leads to an equatorial wind that is an order of magnitude denser than that in the pole. Although the original Lamers and Pauldrach model was for a 1D sector wind, the effect should also operate for 2 D models.

3. The Two Dimensional Models

Here we briefly introduce the "wind compression models". These will be discussed further in the papers by Bjorkman and by Owocki later in this session. A 2 D Magnetic picture is also discusssed, as it introduces another mechanism by which material can be brought from higher latitudes to the star's equatorial region.

3.1. WIND COMPRESSION MODELS

These are the basic ideas underlying the Wind Compression models. As an outflow passes beyond the sonic point the gas pressure gradient becomes of negligible importance in accelerating the outflow. So, for the supersonic portion of a B[e] star wind there would be just two forces operating: the outward radiative acceleration owing to lines, and the inward acceleration of gravity. Unless the rotation of the star is large, both of these forces should be radial. In such a case the outflow obeys conservation of angular momentum. Therefore, we can picture the outflow as being composed of non-interacting parcels of material which are launched at the sonic point and remain in the orbital plane defined at the launch point. Figure 3 shows the trajectories of 3 parcels launched at some latitude with three different initial conditions. There are three parameters that determine the shape of the trajectories. (1) the rotation rate, Ω, (2) the ratio $\zeta = v_\infty/V_{esc}$ of the terminal velocity to the escape speed, (3) the β velocity law index; in which low values (eg. $\beta \sim 0.5$) correspond to laws that accelerate rapidly, while high values (eg. $\beta \sim 2$) correspond to velocities that grow slowly with radius.

Going from trajectory **a** to **c** in figure 3, corresponds to increasing Ω or β, or lowering ζ. In general, the longer the parcel remains near the star, the more the trajectory is wrapped around the star. If a trajectory is like **c**, in wrapping by more than 90 degrees, it is called an "equator crossing" trajectory. Analogous trajectories occur for example for space probes launched

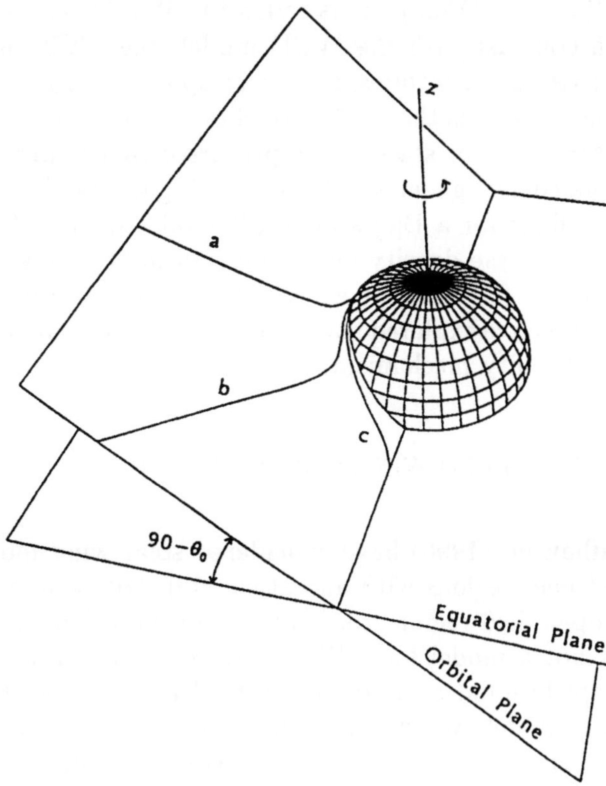

Figure 3. The trajectories ***in the orbital plane*** of wind material originating at co-latidude angle, θ_o. Three orbits shown illustrate an increased "wrapping" of the trajectory around a star. For example, the trajectory c, which reaches the equator, corresponds to the star having a faster rotation or a slower velocity law than is the case for trajectory a. (From Ignace, Cassinelli & Bjorkman 1996)

from a non-equatorial latitude on earth. The orbital plane of a probe is inclined to the earths equator, and hence the trajectory passes through the earth's equatorial plane as the probe leave the earths gravitational field. In the case of stars with winds, the trajectories are associated with a continuously streaming outflows from both the upper and lower hemispheres of the star. Therefore, the trajectories can take the parcels to the equatorial region, but equator *crossing* is not possible because there is a corresponding stream of material from the opposite hemisphere. Therefore collisions and shock compression occur instead. The shock compression is a basic feature of the *Wind Compressed Disk* (WCD) model of Bjorkman and Cassinelli (1993). The nature of the shocked disk was further explored by Owocki, Cranmer and Blondin (1994).

A milder version of the wind compression model has been developed by Ignace, Cassinelli, and Bjorkman (1996) and is called the *Wind Compressed Zone* (WCZ) Model. In contrast with the WCD model, the WCZ model does not have equator crossing trajectories, so it corresponds to stars with somewhat slower rotation rates. It is the WCZ model that is most appropriate for modelling B[e] stars, and for stars in the precursor Red Supergiant phase, as has been discussed by Ignace et al. (1996). Figure 4a shows illustrates the flow stream lines for a B[e] star WCZ model with a $\beta = 3$ velocity law. Figure 4b shows the density enhancements in the equatorial regions that can be produced for different values of the rotation ratio, ω. Cassinelli and Ignace (1997) have explored some consequences of combining WCZ effects with the bi-stability mechanism.

3.2. MAGNETICALLY CHANNELED WIND COMPRESSION

Charbonneau and Hundhausen (1996) have considered solar wind models in which there are closed field regions with an extended distribution of hydrostatic material, and open field regions that channel outflow down to the equatorial zone. This is not a model for a B[e] star, but it illustrates the effects that should be kept in mind as a possibility for B[e] envelope structures. Figure 5 illustrates models which have closed regions out to radius $r = a$. The flow beyond that has an even more pronounced equatorward component than the wind compression models shown in Figure 4a. A possible application of the magnetically channelled models is in explaining the long term V/R variations of H α in Be stars. The fields can also add angular momentum to the wind in a way that is not possible with the wind compressed zone model.

4. Magnetic Field Determinations

We have seen that magnetic fields can lead to several potentially interesting effects in the winds of B[e] and related stars. A chronic problem with magnetic models is that the fields introduce many new parameters, and there is so few observational constraints. The observational situation should change over the next decade. The measurement of the fields in winds is possible by using the Hanle mechanism. The application of the mechanism to stellar astronomy will require observations of the polarization across the wind resonance lines in the far ultraviolet. The Hanle effect for stellar astronomy is described by Ignace, Nordsieck, and Cassinelli, (1997a, b), and by Cassinelli and Ignace (1996).

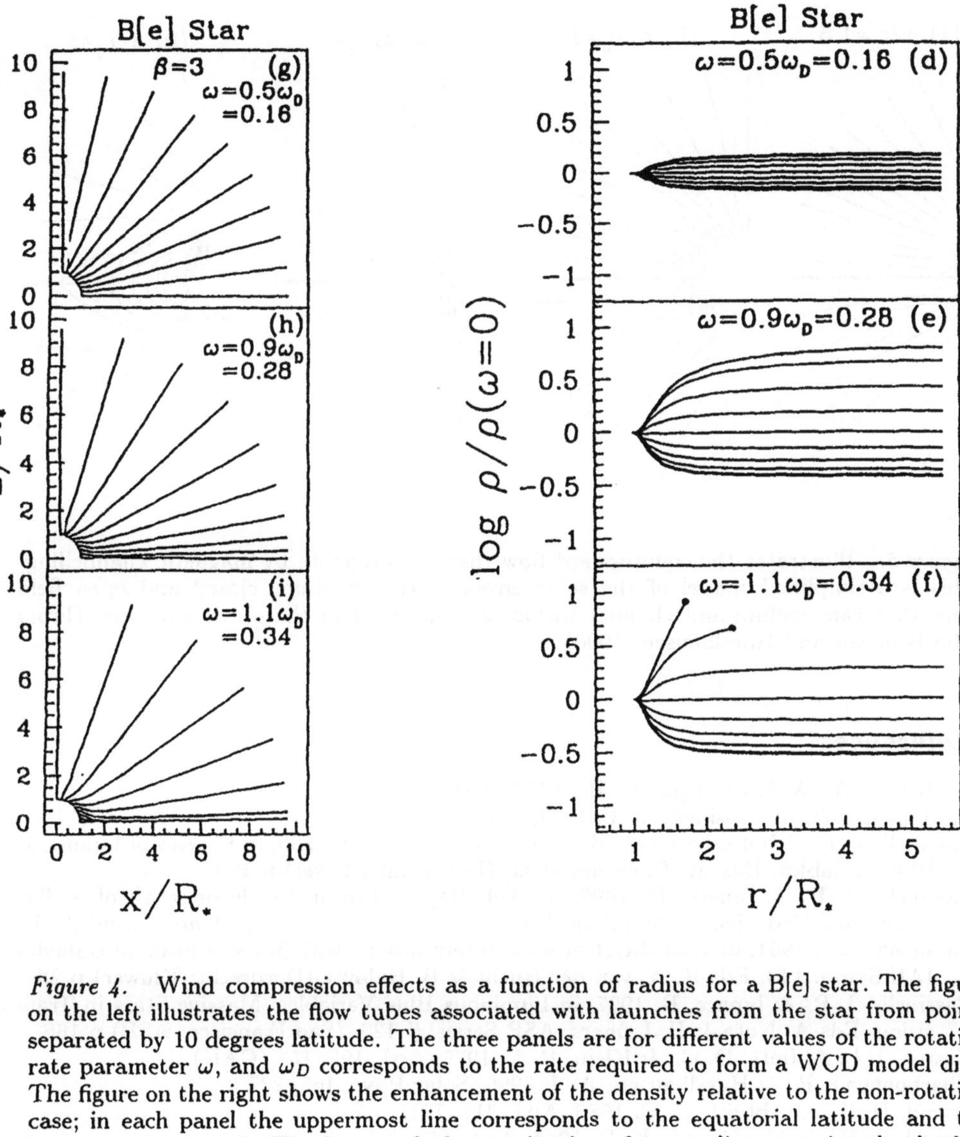

Figure 4. Wind compression effects as a function of radius for a B[e] star. The figure on the left illustrates the flow tubes associated with launches from the star from points separated by 10 degrees latitude. The three panels are for different values of the rotation rate parameter ω, and ω_D corresponds to the rate required to form a WCD model disk. The figure on the right shows the enhancement of the density relative to the non-rotating case; in each panel the uppermost line corresponds to the equatorial latitude and the lowermost to the pole. The dots mark the termination of stream lines entering the shocked disk. (From Ignace, Cassinelli, and Bjorkman 1996)

5. Acknowledgements

I am thankful to Rico Ignace, Jon Bjorkman, Ken Nordsieck, Henny Lamers, Stan Owocki, and John Mathis for useful conversations regarding the various models described here. This research was supported by NSF grant AST 9417186.

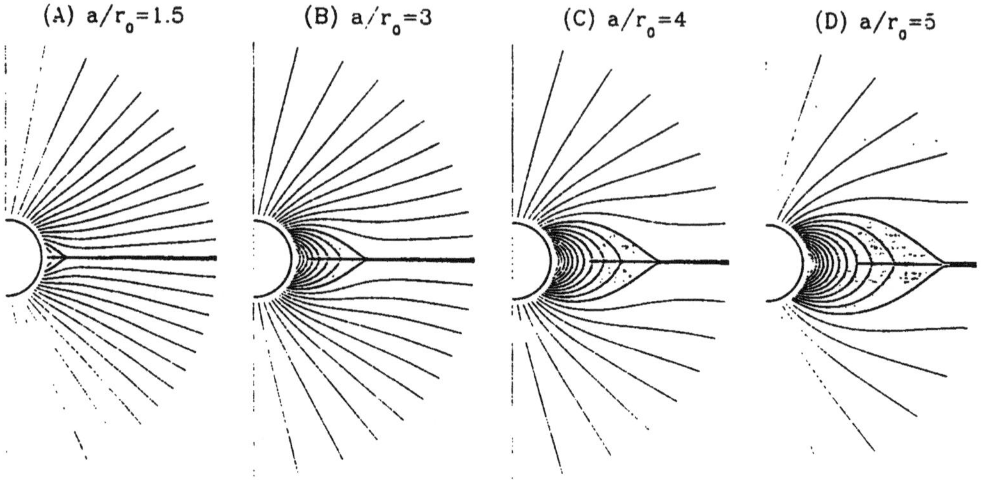

Figure 5. Illustrates the equatorward flow that can originate by magnetic channelling. This is a simplified model of the solar envelope that contains closed and open field lines that can confine and channel the ionized material in the outer envelope. (From Charbonneau and Hundhausen, 1996)

References

Belcher, J. W., & MacGregor, K. B., 1976, ApJ, 268, 498.

Bjorkman, J. E. & Cassinelli, J. P. 1993, ApJ, 409, 429.

Cassinelli, J. P., Schulte-Ladbeck, R., Poe, C. & Abbott, M. 1990, in Physics of Luminous Blue Variables, Eds. K. Davidson et al. (Dordrecht:Kluwer) p 129.

Cassinelli, J. P., & Ignace, R. 1996, in Wolf Rayet Stars in the Framework of Stellar Evolution: 33rd Liege Colloquium. Eds J. M. Vrieux et al. (Liege: Univ. Liege) p 531.

Cassinelli, J. P. 1991, in Wolf Rayet Stars & Interrelations with Massive Stars in Galaxies IAU Symp. 143. Eds K. A. van der Hucht & B. Hidayet (Dordrecht: Kluwer) p 289

Cassinelli, J. P., & Ignace, R. 1997, in Luminous Blue Variables: Massive Stars in Transision. Eds A. Nota & H. Lamers, ASP Series # 120. (San Francisco: ASP) p 166.

Castor, J. I., Abbott, D. C., & Klein, R. I., 1975, ApJ, 195, 157 (CAK).

Charbonneau, P., & Hundhausen, A. J. 1996, Solar Phys. 165, 237.

Friend, D. B., & Abbott, D. C. 1986, ApJ, 311, 701.

Friend, D. B., & MacGregor, K. B. 1984, ApJ, 282, 591.

Hartmann, L. W. & MacGregor K. B. 1982, 259, 180

Ignace, R., Cassinelli, J. P., & Bjorkman, J. E. 1996, ApJ, 459, 671.

Ignace, R., Cassinelli, J. P. & Nordsieck, K, H., 1997, in Luminous Blue Variables: Massive Stars in Transision. Eds. A. Nota & H. Lamers, ASP Series # 120. (ASP: San Francisco) p 198.

Ignace, R., Nordsieck, K. H., & Cassinelli, J. P. 1997, ApJ, 486, 550.

Kudritzki, R. P., Pauldrach, A. W. A., Puls, J., & Abbott, D. C., 1989, A&A, 219, 205.

Lamers, H. J. G. L. M., & Pauldrach, A. W. A. 1991, A&A, 244, L5.

Lee, U., Saio H., & Osaki, Y. 1991, MNRAS, 250, 432.

Maheswaran, M. & Cassinelli, J. P. 1988, ApJ, 335, 931.

Maheswaran, M. & Cassinelli, J. P. 1992, ApJ, 386, 695.

Owocki, S. P., Cranmer, S. R., & Blondin, J. M. 1994, ApJ, 424, 887.

Poe, C. H. & Friend, D. B. 1986, ApJ, 311, 317.
Poe, C. H., Friend, D. B., & Cassinelli, J. P. 1989, ApJ 336, 888.
Pauldrach, A. W. A. & Puls, J., 1990, A&A, 237, 409.
Weber, E. J., & Davis, L. 1967, ApJ, 268, 228.
Zickgraf, F. -J., Wolf, B., Stahl, O., Leitherer, C., Appenzeller, I., A&A, 143, 421.

Discussion

Zickgraf: What velocity contrast between pole and equator can your WCD models produce? Is it a factor of 5 or 10 or only 2 or so?
Cassinelli: In WCZ models there can be a contrast by a factor of about 3 or so. However if a disk forms, the velocity in the disk can be much slower as was shown in the paper by Owocki, Cranmer and Blondin.
Waters: What is the evidence for rotation in the high density equatorial flows of B[e] supergiants? This is a useful discriminant between disks in Be stars (Keplerian) and "disks" in B[e] stars (outflow).
Zickgraf: In a few cases we see double peaked [O I] and Fe II (optical lines). In others we found a dependence of the line width on excitation. We interpreted this in terms of rotational broadening decreasing outwards. The UV spectra of R 50, R 82 and Henize S 22 clearly show outflow in the Fe II resonance lines, i.e. they show normal P Cyg profiles, with blue shifted absorption components. It appears that although indications for rotation in the disks exist, radial outflow seems to be more important.
Jaschek: In this conference we have seen that there exist at least two kinds of B[e] stars, the supergiant ones (like those in the Magellanic Clouds) and the ones near the main sequence. To what kinds apply your models?
Cassinelli: Well, I had the supergiant variety in mind while preparing my talk. The near main sequence objects are presumably more like the original WC disk model for which a shock compression of equatorial material is needed.
Lanz: Why are outflowing disks not Keplerian?
Cassinelli: The stars are rotating at speed far below break up which means their equators are rotating at below Keplerian speed. So the stars would need to supply additional angular momentum to have the material in the disk go at Keplerian speed. In the magnetic rotator model some angular momentum is added to the outflow, but it remains sub-Keplerian. Only the Lee, Saio et al. viscosity driven disk model leads to near Keplerian motion. What is important to realize is that "disks" can arise just from the concentration of the outflow toward the equatorial plane.
Owocki: The non-Keplerian rotation of WCDs is also one of its weaknesses. Because material is not in a stable orbit, the inner disk has infall, while the outer disk is driven outwards. The net result is that no material stays in the disk more than a few days. Thus the WCD model could never explain

very long term variations, like the V/R variations seen in Be stars.
Stee: How and where do you form dust with a WCD model?
Cassinelli: The formation of dust will be discussed in the next presentation, by Jon Bjorkman.
Friedjung: What sort of column density in the equatorial plane can you obtain from your models? Old work with Muratorio on Hen S 22 indicated that even the [Fe II] lines are beginning to be optically thick. This was confirmed more recently by Viotti with other observations. In addition we were able to fit spectral energy distributions of B[e] supergiants by an optically thick disk, irradiated by a central star.
Cassinelli: The winds of supergiant B[e] stars should be nearly optically thick, to electron scattering, even in the polar regions. So if there is a factor of ten enhancement in the equator, the column density through the equatorial region shall be of order 10^{24} - 10^{25} cm^{-2}.
Zickgraf: For three supergiants we have the equatorial wind velocities. They are of the order of 100 $km.s^{-1}$. What we see in these stars is obviously outflow.
Bjorkman: As I will discuss in my talk, the radial electron scattering optical depth of the disk may be $\tau \sim 10$. I suspect this is large enough that the iron lines are optically thick.

WIND-COMPRESSED DISKS AND DUST FORMATION IN B[e] STARS

J.E. BJORKMAN
University of Toledo, Ritter Observatory
Dept. of Physics & Astronomy, Toledo, Ohio 43606, USA

Abstract. We investigate whether the Wind Compressed Disk (WCD) model can reproduce the spectral energy distribution of B[e] stars. In particular, we model the LMC B[e] star R126. Using a Monte Carlo radiative equilibrium and radiation transport code, we calculate the dust condensation radius, the 2-D temperature distribution, and the emergent flux as a function of viewing angle for a WCD model geometry. We find that the WCD model is consistent with the observational data if the dust is composed of hydrogenated amorphous carbon instead of an MRN mixture containing silicates. We also determine a criterion for dust condensation and show that the densities in the WCD outflow are large enough to form dust while spherically symmetric winds cannot form dust. This explains why B[e] stars have dust emission while normal supergiants do not.

1. Introduction

Perhaps the most unusual characteristic of the B[e] class of stars, specifically the LMC supergiants investigated by Zickgraf et al. (1986), is the presence of hot dust emission. It is unusual because no other class of hot stars (other than late WC stars) shows evidence of circumstellar dust in their winds. Based on their similarity to classical Be stars, Zickgraf et al. (1985, 1986) proposed a model for the stellar outflow comprised of a fast wind in the polar regions and a dense equatorial circumstellar disk. Recent HST images of the striking bipolar nebulae around highly evolved stars, such as the η Car homunculus or the Egg nebula, indicate that circumstellar disks are quite common. How such disks are formed is not entirely certain, but Bjorkman & Cassinelli (1993) and Owocki, Cranmer, & Blondin (1994) show how stellar rotation can naturally lead to the formation of a dense

A. M. Hubert and C. Jaschek (eds.), B[e] Stars, 189–204.

wind compressed disk (WCD). More recently Owocki, Cranmer, & Gayley (1996) have questioned whether additional effects that produce non-radial components of the CAK (Castor, Abbott, & Klein 1975) line force can suppress the formation of a disk (see Owocki elsewhere in these proceedings). Whether or not the WCD model applies to B[e] stars depends on the details of the wind driving mechanism, which is not necessarily well understood for B stars.

Subject to this caveat, we apply the WCD model to B[e] stars and calculate the radiative equilibrium temperature of the disk. We find that if the star rotates fast enough to form a WCD, then the disk is dense enough to form dust at large radii (where the radiative equilibrium temperature drops below the dust condensation temperature), and that the WCD model can successfully reproduce the observed spectral energy distribution. This model also explains why dust only forms in B[e] stars and not in other hot supergiant winds; the density of a spherically symmetric wind is too low for dust to form at the condensation radius, but when that same mass loss is redirected through the WCD, the density becomes large enough for dust formation.

2. Rotation and Wind-Compression

In a non-rotating stellar wind, the streamlines are radial, but when the star is rapidly rotating, the fluid elements tend to orbit the star as they are driven outward by the radial forces (see Cassinelli elsewhere in these proceedings). Material from high latitudes orbits toward the equator where it collides with material from the opposite hemisphere. Since the flow velocities are supersonic, this collision results in a pair of shocks above and below the equatorial plane. Between the shocks, the shock compression produces a thin, dense, wind-compressed disk. Thus the primary effect of rotation is not to change the total mass loss rate from the star; instead, it *redistributes* the mass loss such that most of the mass lost from the stellar surface is funneled into the small solid angle of the equatorial disk (typically the opening angle, HWHM, of the disk is only about 3°), which results in the equatorial disk being about 100 times more dense than the polar outflow.

2.1. DISK FORMATION

To form the shock compressed disk, the star must rotate fast enough that the streamlines attempt to cross the equator. Thus there is a minimum stellar rotation speed, above which a disk forms. Below this rotation rate there is still some compression toward the equator, but it only produces a mild density enhancement (typically less than a factor of 10). To distinguish these two classes of equatorial density enhancements, we call the rapidly rotating shock compression disks "WCDs", while the cases with rotation speeds below the disk formation threshold, we call "wind compressed zone" (WCZ) models (Ignace, Cassinelli, & Bjorkman 1996).

2.2. BISTABILITY MECHANISM

Although rotation typically does not appreciably change the mass flux from the stellar surface, there is an important exception that occurs for B[e] stars. This exception is the bistability mechanism discovered by Pauldrach & Puls (1990) in their study of the mass loss from P Cyg. They were studying the effect of lowering the surface gravity (i.e., increasing the stellar radius) on the mass loss rate from the star. They found, for an effective temperature of 19000 K, that when the surface gravity dropped below $\log g_{eff} = 1.6$, there was a sudden increase in the mass flux (about a factor of 3) and a corresponding decrease in the wind terminal speed. This was attributed to the wind becoming optically thick in the photoionizing Lyman continuum, which produces a shift in the ionization balance that changes the number of optically thick driving lines in the wind.

This discontinuity in the terminal speed has now been observed in B supergiants by Lamers, Snow, & Lindholm (1995). They find that earlier than $\log T_{eff} = 4.3$ the data are consistent with $v_\infty/v_{esc} = 2.7$, and later than that v_∞/v_{esc} drops to 1.3. They also find that there may be a second bistability jump at $\log T_{eff} = 4.0$ where v_∞/v_{esc} becomes as small as 0.7.

Since B[e] stars have effective temperatures near the bistability jump, Lamers & Pauldrach (1991) investigated how it may affect the outflow from B[e] stars. They supposed that the star is rapidly rotating and gravity darkened and estimated the hydrogen Lyman column density as a function of latitude. They found that the star could be optically thin in the polar regions and optically thick in the equatorial latitudes. Thus the bistability jump could occur at some mid-latitude, producing denser slower outflow in the equatorial zones, which they argue could be responsible for the B[e] phenomenon. Since the bistability jump is not that large, the density contrast from pole to equator in their models is probably at most about a factor of 10.

Lamers and Pauldrach did not include the 2-D wind compression effects mentioned above. Although we will not attempt to quantitatively combine the bistability mechanism with the wind compressed disk model, it is nonetheless interesting to explore how the two models may interact with each other. If bistability occurs at low latitudes, then the outflow originating at low latitudes has a smaller-than-expected terminal speed and larger mass flux. This implies that the low latitude flow will be more slowly accelerating, which will increase the wind compression effects and lower the threshold stellar rotation speed for forming a disk. On the other hand, wind compression effects increase the low latitude densities, which increases the Lyman column density for two reasons. First, the larger density increases the recombination rates, which lowers the hydrogen ionization fraction and increases the Lyman bound-free opacity. Second, the increased optical depths in the equator prevent the photoionizing flux from reaching the equator and redirect it poleward, since the light will tend to diffuse out the path of least resistance. Decreasing the photoionizing flux also increases the Lyman bound-free opacity, so both these effects imply that the bistability mechanism will occur more easily in the presence of wind-compression effects. Since wind-compression enhances the likelihood of bistability and bistability enhances wind compression, there is now the possibility of some kind of feedback that will enhance the disk density even more than one would naively think. Whether or not such a runaway could occur requires detailed computations that we will not attempt in this paper. Here we will merely use bistability as a mechanism for motivating the possibility that the low latitude outflow may have terminal speeds as small as $v_\infty/v_{esc} = 1.3$.

3. Application to R126

As a specific star to model, we chose R126 as the prototypical B[e] star. It is the brightest B[e] star, which means that it is well observed. It is also believed to be pole-on, $i \sim 0$ (Zickgraf et al. 1985, 1986), which helps simplify our model fitting by reducing ambiguities about the stellar spectral type resulting from circumstellar reddening — pole on stars have the least amount of circumstellar reddening, so their spectral types are easier to ascertain. Zickgraf et al. obtained Johnson 13 color photometry of R126. They found that their spectral energy distribution (SED) indicates $T_{eff} = 22500\,K$ with a radius $R = 72\,R_\odot$ (using the distance of the LMC). From evolutionary tracks they find a zero age main sequence mass $M_{ZAMS} = 80\,M_\odot$ and estimate a current mass $M = 40\,M_\odot$. The SED also shows a near IR excess, beginning at V and extending to H (see fig. 3) with a slope characteristic of free-free emission and a density falling as r^{-2}; i.e., constant velocity expansion. Beginning at K, there is dust emission that peaks around L, indicative of hot dust ($T_{dust} \sim 900\,K$).

We now wish to investigate whether a WCD model can reproduce this SED. We suppose that the free-free emission originates in the inner regions of the WCD where the gas is completely ionized. This disk material flows outward, and at the point where the dust radiative equilibrium temperature equals the dust evaporation temperature, dust can condense. Figure 1 illustrates this geometry.

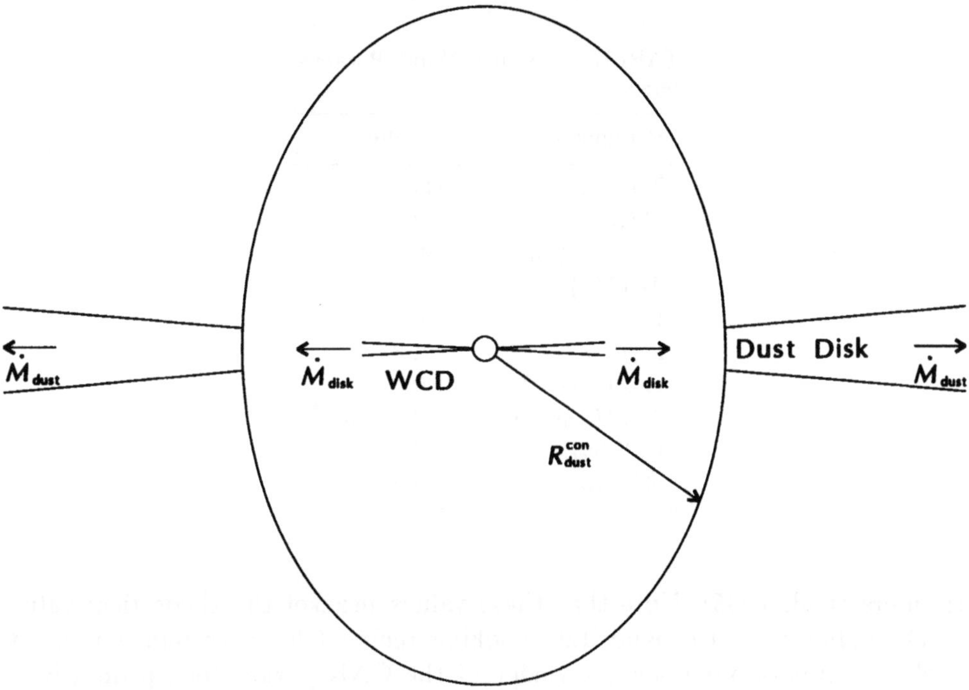

Figure 1. Circumstellar Geometry. The star is surrounded by a thin equatorial wind compressed disk. At small radii this disk is fully ionized and free-free emission is responsible for producing the near IR excess. The material in the inner disk continues to flow outward, and at the dust condensation radius, R^{con}_{dust}, dust forms in the outer disk. This dust is heated by the stellar radiation field to the radiative equilibrium temperature. It then reemits the absorbed starlight via thermal dust emission at long wavelengths, producing the far IR excess.

3.1. STELLAR WIND PARAMETERS

In the wind compressed disk model, some fraction of the total stellar mass loss enters the disk, depending on the stellar rotation speed. To estimate the stellar mass loss rate, we use Kudritzki's "cooking recipe" (Kudritzki et al. 1989). We adopt CAK parameters $k = 0.159$, $\alpha = 0.60$, and $\delta = 0.118$ (Bjorkman & Cassinelli 1993), which were based on the values computed by Abbott (1982). Using these parameters, we find a theoretical stellar mass loss rate $\dot{M} = 4.6 \times 10^{-5}\, M_{\odot}\, yr^{-1}$ (see Table 1 for a complete list of the adopted stellar wind parameters). Interestingly, $\log g_{eff} = 1.8$. Since the bistability jump occurs at $\log g_{eff} \approx 1.6$, we find it likely that the bistability mechanism could be operating at low latitudes where the centrifugal force reduces g_{eff} below the bistability limit. For this reason, we adopt $v_{\infty}/v_{esc} = 2.7$ for high latitudes and $v_{\infty}/v_{esc} = 1.3$ for low latitudes, which are the observed values for the terminal speeds on either side of the bistability jump

TABLE 1. Stellar Wind Parameters

Parameter	Value
T_{eff} (K)	22500
R ($R_\odot$)	72
M_{ZAMS} ($M_\odot$)	80
M ($M_\odot$)	40
Γ	0.7
$\log g_{eff}$	1.8
v_{esc} (km s^{-1})	240
$\dot{M}$ ($M_\odot\, yr^{-1}$)	4.6×10^{-5}
v_∞^{pole}/v_{esc}	2.7
v_∞^{eq}/v_{esc}	1.3

(Lamers et al. 1995). Note that these values bracket the theoretical value (1.7) predicted by the Kudritzki cooking recipe (the discrepancy is most likely a result of our poor knowledge of the CAK parameters, primarily α — the bistability jump implies a discontinuous change, which cannot be obtained from Abbott's optically thin theoretical calculations).

Since the disk is formed from the low latitude outflow, we use the low latitude terminal speed, $v_\infty/v_{esc} = 1.3$, to determine the disk formation threshold,

$$\left(\frac{V_{rot}}{V_{crit}}\right)_{th} = \begin{cases} 0.3 & (\beta = 3), \\ 0.8 & (\beta = 0.8), \end{cases} \tag{1}$$

where β is the velocity law exponent (i.e., $v = v_0 + [v_\infty - v_0][1 - R/r]^\beta$). Since we do not know how rapidly a bistable wind accelerates, we have chosen both a slow ($\beta = 3$) and a fast ($\beta = 0.8$) acceleration to illustrate the possible range. Using the escape speed in Table 1 to determine the critical speed $V_{crit} = V_{esc}/\sqrt{3}$, the required rotation speed to form a WCD is

$$(V_{rot})_{th} = \begin{cases} 40\, km\, s^{-1} & (\beta = 3), \\ 110\, km\, s^{-1} & (\beta = 0.8). \end{cases} \tag{2}$$

These values imply that it is relatively easy to form a disk, especially if the wind is slowly accelerating ($\beta = 3$).

3.2. INNER DISK: FREE-FREE EXCESS

The stellar rotation rate determines what fraction of the stellar mass loss goes into the disk. Since B[e] stars have so much envelope emission, photospheric lines are not observable in the optical, so it is difficult to directly measure $v \sin i$ values. Therefore the stellar rotation speed (or equivalently the disk density) is a free parameter. We can directly determine the disk density by fitting the free-free (plus bound-free) emission that produces the near IR excess (see Waters 1986 for details). For convenience, we assume an isothermal disk with $T_{disk} = 15000\,K$. This temperature is motivated by the radiative equilibrium temperatures estimated by Millar and Marlborough (1997) for classical Be star disks, which like B[e] stars also have low ionization species like Fe II in their disks (Zickgraf et al 1985, 1986). We parameterize the density by

$$\rho = \rho_0 \left(\frac{R}{r}\right)^n \sin^m \theta \quad . \tag{3}$$

The radial exponent $n = 2$ is determined by the slope of the free-free emission (Wright and Barlow 1975), and the WCD model predicts a disk opening angle $\Delta\theta_{WCD} = 3°$ (Owocki et al. 1994), corresponding to $m = 500$. The amplitude of the free-free excess then determines

$$\rho_0 = 8 \times 10^{-12}\, g\, cm^{-3} \quad . \tag{4}$$

Note that this density produces a radial electron scattering optical depth $\tau_{es} = 13$, so the disk is quite optically thick in the radial direction. Comparing this density to the stellar mass loss rate (via the continuity equation), we find $\rho_{eq}/\rho_{pole} \approx 80$, which implies that we must have a WCD (as opposed to a WCZ model). Thus the star must be rotating faster than the disk formation threshold. If we know the outflow velocity in the disk, we can convert this density to a mass loss rate through the disk. Zickgraf et al. (1996) find that typical outflow velocities in the disks of edge-on B[e] stars (like S22) are about $v_\infty^{disk} \sim 75\, km\, s^{-1}$. Using this disk terminal speed, we find that the mass loss through the disk must be

$$\dot{M}_{disk} = 4\pi R^2 \sin \Delta\theta_{WCD} \rho_0 v_\infty^{disk} = 1.6 \times 10^{-5}\, M_\odot\, yr^{-1} \quad . \tag{5}$$

Thus the fraction of the total mass loss that enters the disk is $\dot{M}_{disk}/\dot{M} = 0.3$, which according to the WCD model, would require a stellar rotation rate

$$\left(\frac{V_{rot}}{V_{crit}}\right) = \begin{cases} 0.37 & (\beta = 3), \\ 0.85 & (\beta = 0.8), \end{cases} \tag{6}$$

or equivalently

$$V_{rot} = \begin{cases} 50\, km\, s^{-1} & (\beta = 3), \\ 120\, km\, s^{-1} & (\beta = 0.8). \end{cases} \tag{7}$$

3.3. OUTER DISK: THERMAL DUST EMISSION

The mass loss through the inner disk continues outward and at the dust condensation radius R_{dust}^{con}, dust forms (see Fig. 1). We could use the continuity equation to determine the total density (gas plus dust) in the outer disk; however, we do not know the dust formation efficiency. So again we prefer to use a parameterized density distribution to fit the thermal dust emission. The radiation transfer code we use employs a 3-D Monte Carlo simulation technique that calculates the dust radiative equilibrium temperature as a function of position, as well as the emergent spectrum accounting for dust scattering, absorption, and thermal emission; it is fully valid for optically thick calculations (Bjorkman & Wood 1998). For computational efficiency, we choose an ellipsoidal parameterization for the dust opacity

$$\kappa_\nu \rho = \frac{\kappa_\nu \rho_0^d R^2}{x^2 + y^2 + (a/b)^2 z^2} \quad (r > R_{dust}^{con}), \tag{8}$$

where κ_ν is the dust opacity per (total) mass, and (a/b) is the ratio of semi-major to semi-minor axes. To approximately match the pole to equator density ratio, we choose $(a/b) = 10$, which corresponds to a disk opening angle $\Delta\theta_{dust} = 6°$; this is somewhat larger than the inner disk, but we feel it is more important to match the pole-to-equator density ratio than the exact opening angle of the disk.

To perform the dust temperature and emergent flux calculations, we use the Monte Carlo code to inject a stellar spectrum (modified by the electron scattering and free-free plus bound-free emission from the inner disk) into the dust opacity distribution. Note that since $R_{dust}^{con} \gg R$, the star plus inner disk is essentially a point source, but since the inner disk is optically thick, this input spectrum is not isotropic; the polar flux is about three times larger than the equatorial flux. We then vary R_{dust}^{con}, which implicitly determines the dust condensation temperature, and ρ_0^d (or equivalently τ_V, the V-band equatorial radial optical depth). Varying R_{dust}^{con} changes the wavelength of the dust emission peak (because the maximum dust temperature is determined by R_{dust}^{con}), and the amplitude of the dust emission determines τ_V.

Fitting the observed flux, we find

$$R_{dust}^{con} = 2000R = 700\,AU, \tag{9}$$

which yields a dust condensation temperature $T_{dust}^{con} \approx 700\,K$. Figure 2 shows the dust temperature as a function of radius for various values of the co-latitude θ. Note that the temperature in the pole follows the optically thin approximation, $T \propto (R/r)^{0.4}$, while in the equator, the temperature is initially equal to the optically thin approximation, but drops faster owing to

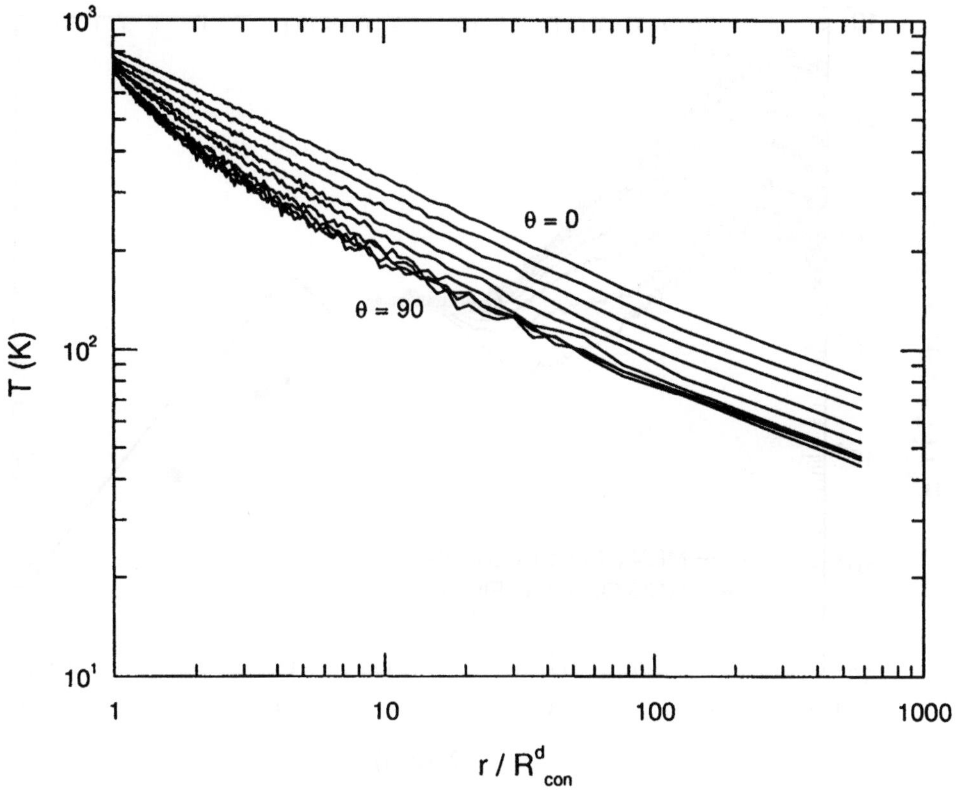

Figure 2. Dust Radiative Equilibrium Temperature. Shown is the dust temperature as a function of radius (normalized to the dust condensation radius) for several values of the co-latitude θ.

the optically thick shielding by the disk. Thus we see that self-shielding by an optically thick disk does not appreciably change the dust condensation radius (because the leading edge of the disk heats up to the optically thin temperature); it merely reduces the temperature in the outer portions of the disk.

Figures 3 and 4 show the emergent flux for two different dust opacities: an MRN mixture (Mathis, Rumpl, & Nordsieck 1977), and hydrogenated amorphous carbon (HAC), both with a power law distribution of grain sizes. Interestingly the MRN mixture, which contains silicate grains, does

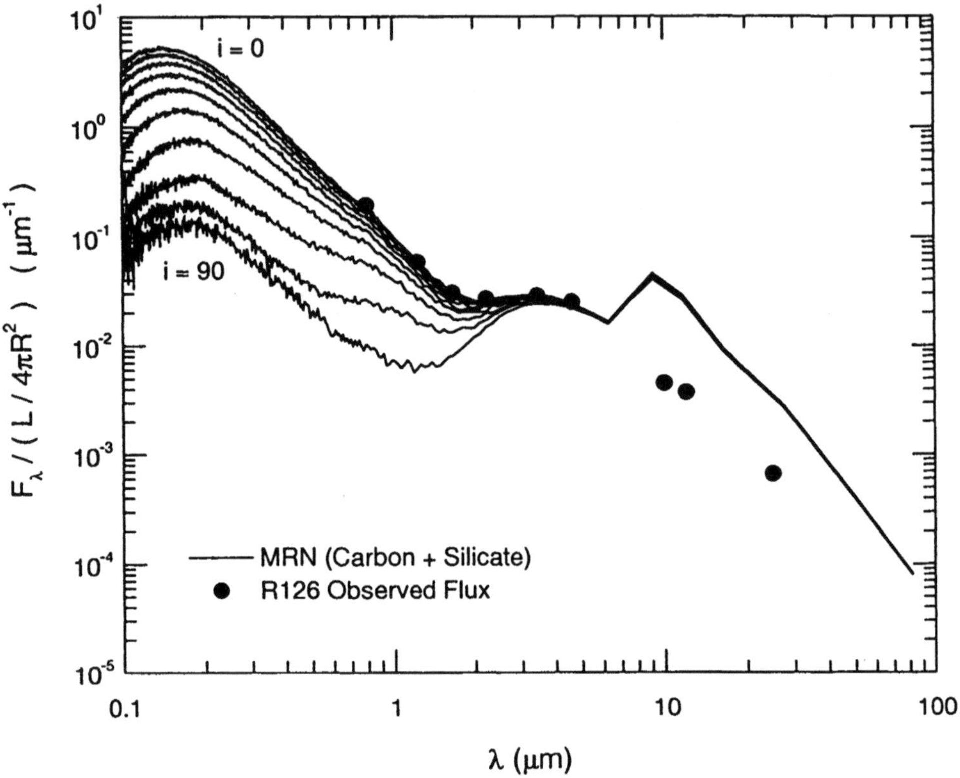

Figure 3. Spectral Energy Distribution for MRN Dust. Shown is the emergent flux as a function of inclination angle i. At long wavelengths the dusty disk is optically thin, so the emergent flux is independent of inclination angle, but when the disk is optically thick (at short wavelengths) large intrinsic reddening occurs for the edge-on cases ($i = 90°$). Note the discrepancy between the model calculations (solid lines) and the observed flux (dots) at long wavelengths and at the 10 μm silicate emission feature.

not match the far IR flux, and high resolution data do not show any evidence of the 10 μm silicate emission feature (Roche, Aitken, & Smith 1993). One might imagine that changing the particle size distribution of the MRN mixture might help remove the discrepancy; however, since the 10 μm silicate emission is produced by the Si-O stretch, it is probably difficult to remove the silicate emission without removing all the silicate particles (Mathis, private communication). On the other hand, the HAC opacity matches both the mid- and far-IR fluxes much better than the MRN mixture. The simplest explanation is that the stellar outflow is carbon rich with $C/O > 1$.

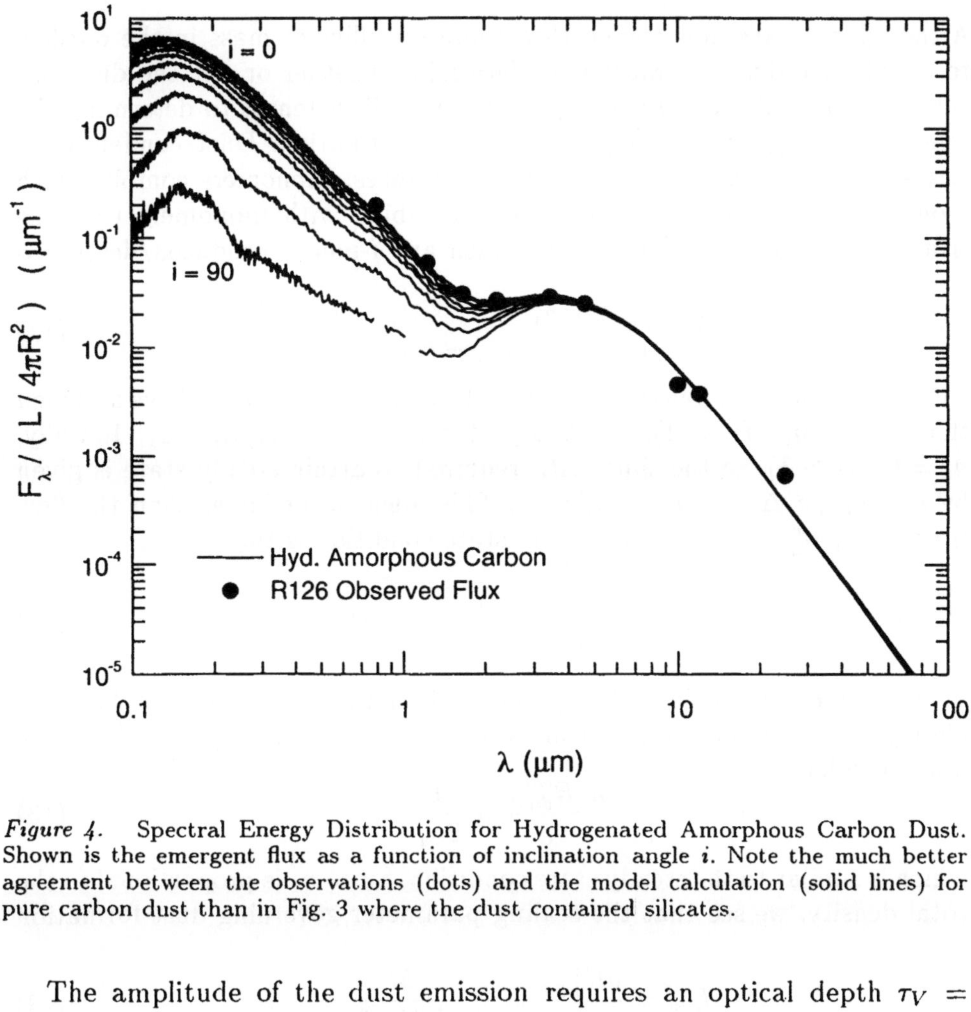

Figure 4. Spectral Energy Distribution for Hydrogenated Amorphous Carbon Dust. Shown is the emergent flux as a function of inclination angle *i*. Note the much better agreement between the observations (dots) and the model calculation (solid lines) for pure carbon dust than in Fig. 3 where the dust contained silicates.

The amplitude of the dust emission requires an optical depth $\tau_V = 2.5$ for the HAC dust. Assuming a dust to gas ratio of about 200, $\kappa_V = 300\, cm^2\, g^{-1}$. Using the continuity equation (5) to compare the mass loss rate through the dust disk to the inner disk (which has $\tau_{es} = 13$ and $\sigma_e = 0.3\, cm^2\, g^{-1}$), and assuming the outflow velocity of the dust is the same as in the inner disk, we find

$$\frac{\dot{M}_{dust}}{\dot{M}_{disk}} = \frac{\tau_V \sin \Delta\theta_{dust} \sigma_e R^{con}_{dust}}{\tau_{es} \sin \Delta\theta_{WCD} \kappa_V R} = 0.7 \quad . \tag{10}$$

This implies that the mass loss through the WCD (inner disk) is capable of supplying the required amount of dust if the dust condensation is fairly efficient.

4. Dust Condensation

Although we have now shown that there is sufficient mass in the outflow to supply the dust, we must now determine whether or not the dust can actually form. To address this question, we will attempt to determine the relevant scaling parameter governing dust formation. Dust condensation is a series of two body recombination processes. Monomers combine with monomers to produce dimers; dimers combine with monomers to make trimers, and so forth. The rate at which any i-mer is produced/destroyed is

$$\frac{dn_i}{dt} = \begin{cases} -2\alpha_1 n_1^2 & (i=1), \\ n_1(\alpha_{i-1}n_{i-1} - \alpha_i n_i) & (i>1), \end{cases} \tag{11}$$

where α_i is the recombination coefficient for the i-mer, and we have assumed that $n_1 \gg n_i \quad (i > 1)$. In steady state, $n_i = (\alpha_{i-1}/\alpha_i)n_{i-1}$. Initially, $n_i = 0 \quad (i > 1)$, so the time, Δt_i, required to attain steady state is given by $\alpha_{i-1}n_{i-1}n_1\Delta t_i = (\alpha_{i-1}/\alpha_i)n_{i-1}$. This must occur faster than the flow time scale, R_{dust}^{con}/v_∞, so the steady state condition is that

$$\frac{n_1 R_{dust}^{con}}{v_\infty} > \frac{1}{\alpha_i} \quad (i > 1). \tag{12}$$

To complete the condensation process, the parent $(i = 1)$ must be completely consumed on the flow time scale, so $n_1 < 2\alpha_1 n_1^2 R_{dust}^{con}/v_\infty$. Thus we must also have

$$\frac{n_1 R_{dust}^{con}}{v_\infty} > \frac{1}{2\alpha_1} \quad , \tag{13}$$

which is similar to the steady state condition. Since n_1 is proportional to the total density, we see that the scaling parameter governing dust formation is

$$f_{con} = \frac{\rho R_{dust}^{con}}{v_\infty} \propto \frac{\dot{M}}{\sin\Delta\theta R_{dust}^{con} v_\infty^2} \quad . \tag{14}$$

To determine whether or not dust can form in the WCD outflow, we compare to AGB stars where we know dust forms. The smallest AGB mass loss rate where dust forms is $\dot{M}_{AGB} = 10^{-7}\, M_\odot\, yr^{-1}$, with a stellar radius $R_{AGB} = 100\, R_\odot$. A typical wind terminal speed is $v_\infty^{AGB} = 30\, km\, s^{-1}$ and the dust condensation radius is about $R_{con}^{AGB} = 3R_{AGB}$. Comparing the dust condensation parameter for the AGB star to our B[e] model, we find

$$\frac{f_{con}^{B[e]}}{f_{con}^{AGB}} = \frac{1}{\sin\Delta\theta_{dust}} \left(\frac{\dot{M}_{B[e]}}{\dot{M}_{AGB}}\right) \left(\frac{R_{con}^{AGB}}{R_{con}^{B[e]}}\right) \left(\frac{v_\infty^{AGB}}{v_\infty^{B[e]}}\right)^2 \sim 1 \quad . \tag{15}$$

Since the ability to form dust in our WCD outflow is the same as that in the AGB star, we conclude that dust does indeed condense.

We can also now answer why dust does not form in spherically symmetric supergiant winds. Comparing the WCD outflow to a spherically symmetric model we see that

$$\frac{f_{con}^{WCD}}{f_{con}^{sph}} = \left(\frac{\rho^{WCD}}{\rho^{sph}}\right)\left(\frac{v_{\infty}^{sph}}{v_{\infty}^{WCD}}\right) \sim 10^3 \quad . \tag{16}$$

Therefore a spherically symmetric model has a factor of 10^3 less condensation than a WCD model. Thus the presence of a geometrically thin dense disk is why B[e] stars form dust and normal supergiants do not; it has nothing to do with the ability of the disk to shield itself from the stellar heating — it is simply that the density of a spherical supergiant wind is too small for dust to form at the very large condensation radius.

5. Conclusions

We have shown that the WCD model can successfully reproduce the observed spectral energy distribution of the B[e] star R126. Assuming a theoretical mass loss rate $\dot{M} = 5\times10^{-5}\, M_{\odot}\, yr^{-1}$ and assuming that the bistability mechanism operates at low latitudes ($v_{\infty}/v_{esc} = 1.3$), then the required stellar rotation speed is $50\,\mathrm{km\,s^{-1}}$ for a slowly accelerating wind ($\beta = 3$) and $120\,\mathrm{km\,s^{-1}}$ for a rapidly accelerating wind ($\beta = 0.8$). We have also calculated the 2-D radiative equilibrium temperature distribution of the dust as well as the emergent flux for both an MRN dust mixture and hydrogenated amorphous carbon. The observations do not fit the $10\,\mu$m silicate emission produced by the MRN mixture, and the MRN dust also produces too much far IR emission. On the other hand, if the dust were composed of HAC, then there is no silicate emission, and the far IR fluxes agree with the model calculations. The simplest explanation is that the dust is produced by carbon rich material with $C/O > 1$, which is a bit of a puzzle. Finally we show that the ability to form dust depends on the value $f_{con} = \rho R_{dust}^{con}/v_{\infty}$. Comparing B[e] stars to AGB stars, we find that they have comparable ability to form dust, if the B[e] star has a dense equatorial disk consistent with that produced by the WCD model. Similarly the ability to form dust in a WCD outflow is 1000 times larger than that of a spherical outflow. This explains why normal supergiants do not produce dust, while those with dense equatorial disks, the B[e] stars, do form dust.

References

Abbott, D.C. 1982, ApJ, 259, 282

Bjorkman, J.E., & Cassinelli, J.P. 1993, ApJ, 403, 429
Bjorkman, J.E., & Wood, K. 1998, ApJ, in preparation
Castor, J.I., Abbott, D.C., & Klein, R.I. 1975, ApJ, 195, 157 (CAK)
Ignace, R., Cassinelli, J.P., & Bjorkman, J.E. 1996, ApJ, 459, 671
Kudritzki, R.P., Pauldrach, A.W.A., Puls, J., & Abbott, D.C. 1989, A&A, 219, 205
Lamers, H.J.G.L.M., & Pauldrach, A.W.A. 1991, A&A, 244, L5
Lamers, H.J.G.L.M., Snow, T.P., & Lindholm, D.M. 1995, ApJ, 455, 269
Mathis, J.S., Rumpl, W, & Nordsieck, K.H. 1977, ApJ, 217, 425
Millar, C.E., & Marlborough, J.M. 1997, ApJ, (in press)
Owocki, S.P., Cranmer, S.R., & Blondin, J.M. 1994, ApJ, 424, 887
Owocki, S.P., Cranmer, S.R., & Gayley, K.G. 1996, ApJ, 472, L115
Pauldrach, A.W.A., & Puls, J. 1990, A&A, 237, 409
Roche, P.F., Aitken, D.K., & Smith, C.H. 1993, MNRAS, 262, 301
Zickgraf, F.-J., Wolf, B., Stahl, O., Leitherer, C., & Klare, G. 1985, A&A, 143, 421
Zickgraf, F.-J., Wolf, B., Stahl, O., Leitherer, C., & Appenzeller, I. 1986, A&A, 163, 119
Zickgraf, F.-J., Humphreys, R.M., Lamers, H.J.G.L.M., Smolinski, J., Wolf, B., & Stahl, O. 1996, A&A, 315, 510
Waters, L.B.F.M. 1986, A&A, 162, 121
Wright, A.E., & Barlow, M.J. 1975, MNRAS, 170, 41

Discussion

Lamers: The dust formation radius in your model depends on two factors a) where does the wind/disk recombine b) what is the dust condensation temperature. The dust will form where the wind/disk has recombined and the temperature is smaller than T_c. So I would expect that the dust temperature might be very sensitive to the stellar parameters? Is that observed?
Bjorkman: The dust radiative equilibrium temperature's $T_d = T_*(w)^{0.2}$ (w is the dilution factor) . So the dust temperature depends linearly on the stellar temperature, but only weakly on the stellar equatorial flux, as compared to the spherical flux. Recombination rates are proportional to $1/\sqrt{T}$ so the recombination is weakly dependent on temperature.
Zickgraf: For the mass loss estimate from the Fe II lines you mentioned ($\dot{M} \approx 10^{-7} M_\odot.\mathrm{yr}^{-1}$, we used only a lower limit for the optical depth of the lines, i.e. $\dot{M}$ is a lower limit. In addition the number 10^{-7} contains the ionization factor which we do not know.
Bjorkman: I must have misread your paper. Good. It fixes my problem if the ionization fraction $q \sim 10^{-2}$.
Waters: It is remarkable that the dust condensation temperature is rather low for amorphous carbon in the model fit to R 126. $T_c \approx 800$ K would be more appropriate for silicates. The lack of strong silicate emission in the observed spectrum does not necessarily imply that the dust is C-rich. The free-free slope corresponding to r^{-2} density gradient is quite different from those of classical Be stars (between r^{-3} and r^{-4}) !
Bjorkman: When we fit the spectral energy distribution, we only tried two chemical compositions (MR and hydrogenated amorphous carbon) with a

MR size distribution. It would be very interesting to see if we adjusted the particle size distribution whether or not we could remove the strong 10 μm silicate emission while retaining silicates for the chemical composition. If so this would possibly explain the low dust condensation temperature and no longer imply that the stellar outflow has C/O > 1.

Stee: Do you think that the shock in the WCD model can change the ionization/density structure?

Bjorkman: The WCD shock is relatively weak (< 100 km.s^{-1} typically) and its X-ray luminosity is less than that of the X-rays produced elsewhere in the wind (possibly owing to the wind instability shocks), so I doubt the WCD shock will change the ionization state of the wind. However locally, the shock temperatures may be large enough to collisionally produce supe-rionization (C IV and Si IV) locally in the post-shock material.

Schulte-Ladbeck: You used the WCD model to compare the physical properties of B supergiants and B[e] stars to infer favorable conditions for the formation of the dust around B[e] stars. How does this comparison work for Be stars and the lower luminosity B stars? As far as I know Be stars do not show IR excesses due to dust. What is different in their disks?

Bjorkman: The mass loss rates of classical Be stars are a factor of 10^{-4} smaller than that of R 126. This should reduce the density of the disk below that required for dust formation.

Viotti: Temperature and density are the principal ingredients to help dust formation. The matter temperature can be reduced by the opacity increase especially if there are large density fluctuations, while the density can be increased if "blobs" are formed as a consequence of turbulence in the flowing matter.This process can be effective in mass losing stars without evidence of dense disks. In the models, the turbulence effects have been taken into account?

Bjorkman: We did not include any clumping in our model. I agree that such density enhancements will increase the ability to form dust.

Owocki: (comment regarding the WCD shock rays): The equatorward flow speed in a WCD shock is generally limited to about 25 % of the stars equatorial speed, due to projection effects. Thus in the case of WCD's from slowly rotating stars, that are possible because of high β or low escape velocity V_∞, the WCD shocks must be quite weak, with a shock speed jump $\Delta V \leq 50$ km.s^{-1}, which is too low to provide many X-rays.

Friedjung: Firstly I wish to emphasize that optical Fe II lines of B[e] supergiants have large optical depths and those of the UV 0 volt lines will be much larger than in the Zickgraf et al (1966) paper. Therefore I think that their mass loss rate was underestimated and that your value is more probably correct. Secondly your disk is starting to become optically thick in the optical with respect to electron scattering perpendicular to the disk.

You will not have just optically thin free free emission, but also scattered light from the central star by the disk, which we considered in our old work (Muratorio and Friedjung 1988).

Bjorkman: Since we included both scattering and absorption in our Monte-Carlo radiation transfer, our models should also contain those effects.

LATITUDINAL DEPENDENCE OF RADIATIVELY DRIVEN MASS LOSS FROM RAPIDLY ROTATING HOT-STARS

S. P. OWOCKI

Bartol Research Institute, University of Delaware, Newark, DE 19716

S. R. CRANMER

Harvard-Smithsonian Center for Astrophysics, Cambridge, MA 02138

AND

K. G. GAYLEY

Department of Physics and Astronomy, University of Iowa, Iowa City, IA 50011

Abstract. We investigate the latitudinal variation of radiatively driven mass loss from rapidly rotating hot-stars. Previous analyses have assumed a uniformly bright stellar surface and concluded that the wind mass flux and density should increase with the increased centrifugal force toward the wind *equator*. In contrast, we show here that a gravity darkening in which the surface radiation flux scales with the effective (centrifugally reduced) gravity leads to a dramatically different wind morphology, with the strongest mass loss now occurring toward the relatively bright *poles*. We also review recent work that indicates nonradial (poleward) components of the line-driving force in such rotating winds can effectively inhibit the equatorward wind deflection needed to form an equatorial wind-compressed disk. Finally, we examine the equatorial bistability model, and show that a sufficiently strong jump in wind driving parameters can, in principle, overcome the effect of reduced radiative driving flux, thus still allowing moderate enhancements in density in an equatorial, bistability zone wind.

A. M. Hubert and C. Jaschek (eds.), B[e] Stars, 205–220.

1. Introduction

Several observational reviews of this workshop (e.g. Zickgraf) have emphasized the various lines of evidence that the circumstellar regions around B[e] stars have an enhanced equatorial density, possibly even a circumstellar disk. A key question is whether this enhanced equatorial density might be a direct result of rotational effects that could influence the latitudinal variation of the radiatively driven stellar wind expected from such a luminous B-star. Previous theoretical analyses of winds from rotating hot stars have examined various effects that can increase the equatorial density, including:

- enhanced mass flux and reduced wind speed associated with the centrifugally reduced effective gravity near the equator (Friend and Abbott 1986; Pauldrach, Puls, and Kudritzki 1986);
- Wind Compressed Disks (WCDs) (Bjorkman & Cassinelli 1993; Owocki, Cranmer, and Blondin 1994; Ignace, Cassinelli, and Bjorkman 1996); or
- enhanced mass loss from a "bi-stability zone" triggered by the reduced radiation temperature near the equator (Lamers and Pauldrach 1991).

All these previous analyses have implicitly assumed that the radiative flux is uniformly distributed over the stellar surface. However, to the extent that radiation dominates the energy transport in the envelopes of such hot stars, the classical analysis of von Zeipel (1924) implies a "gravity-darkening" effect, in which the radiative flux from near the equator is reduced in proportion to the centrifugally reduced effective gravity there. Here we review results from recently developed 2-D wind simulations that incorporate the effects of such gravity darkening, as well as nonradial components of the line-driving force (Owocki, Cranmer, and Gayley 1996). As discussed below, both effects turn out to have a surprisingly strong impact on the latitudinal mass distribution of a stellar wind. To provide a basis for understanding these unexpected results, let us first review scaling relations predicted from 1-D models of line-driven winds.

2. Scaling Laws from 1-D Models

Initial investigations (Friend & Abbott 1986; Pauldrach et al. 1986) of the possible role of rotation on radiatively driven winds derived 1-D models based on the standard line-driving formalism of Castor, Abbott, and Klein (1975; hereafter CAK), but now adding the effect of an outward centrifugal acceleration in the equatorial plane, $g_{cent}(r) = V_{rot}^2 R^2/r^3$, where V_{rot} is the equatorial rotation speed at the stellar surface radius $r = R$. Although this centrifugal term declines faster with radius than gravity, the mass loss is fixed at a critical point quite near the stellar surface. This suggests that the effect on the local mass flux at any colatitude θ of a rotating star can be written in terms of a centrifugally reduced, *effective* surface gravity

$$g_{eff}(\theta) = \frac{GM}{R^2}\left(1 - \Omega \sin^2\theta\right), \tag{1}$$

where G and M are the gravitation constant and stellar mass, and $\Omega \equiv V_{rot}^2 R/GM$. We thus rewrite the standard CAK mass loss rate scaling law (i.e. CAK eqn. (46)) in terms of *surface* values of the mass flux $\dot{m} = \rho v$, radiative flux F, and effective gravity g_{eff}, relative to corresponding polar ($\theta = 0$) values $\dot{m}_o$, F_o, and $g_o = GM/R^2$,

$$\frac{\dot{m}(\theta)}{\dot{m}_o} = \left[\frac{F(\theta)}{F_o}\right]^{1/\alpha}\left[\frac{g_{eff}(\theta)}{g_o}\right]^{1-1/\alpha}, \tag{2}$$

where α is the usual CAK exponent, assumed here to be constant at all latitudes (cf. §4.2 below).

For example, if we take $F(\theta) = F_o$, then we obtain the scaling

$$\frac{\dot{m}(\theta)}{\dot{m}_o} = \left[1 - \Omega \sin^2\theta\right]^{1-1/\alpha} \quad ; \quad F(\theta) = F_o\,, \tag{3}$$

which, for $\sin\theta = 1$ and $\alpha \approx 0.6$, provides a reasonably good fit to, e.g., numerical results plotted in fig. 4 of Friend & Abbott 1986.

Since $\alpha < 1$, the exponent $1 - 1/\alpha$ is negative, implying that the mass flux increases monotonically from pole ($\sin\theta \to 0$) toward the equator ($\sin\theta \to 1$). On the other hand, if we apply the standard von Zeipel(1924) (see also Kippenhahn & Weigert 1990) gravity darkening law that the surface flux itself varies in proportion to the centrifugally reduced effective gravity, $F(\theta) \sim g_{eff}$, we obtain

$$\frac{\dot{m}(\theta)}{\dot{m}_o} = 1 - \Omega \sin^2\theta \quad ; \quad F(\theta) \sim g_{eff}(\theta) \tag{4}$$

so that the mass flux now *decreases* towards the equator, with a maximum at the *pole* (see §4)!

Friend and Abbott (1986) likewise find that the terminal wind speed is decreased by rotation in the equator. We can also approximate this result within our 1-D concept of a centrifugally reduced gravity to predict a latitudinally varying wind terminal speed that scales with an effective escape speed, $v_\infty(\theta) \sim v_{esc}\sqrt{1 - \Omega \sin^2\theta}$. The latitudinal variation of density is then obtained from $\rho \sim \dot{m}/v_\infty$.

3. 2-D Dynamical Simulations of Rotating Winds

3.1. EQUATORIAL WIND COMPRESSED DISKS

A major advance in modelling rotating hot-star winds was development of the elegantly simple "Wind Compressed Disk" (WCD) paradigm by Bjorkman & Cassinelli (1993). They noted that, like satellites launched into earth orbit, parcels of gas gradually driven radially outward from a rapidly rotating star should remain in a tilted 'orbital plane' that brings them over the equator. As wind parcels from opposite hemispheres collide over the equator, they form a disk of compressed gas. A key simplification here is to assume that, like gravity, the radiative driving is a radially directed, *central* force, so that the total angular momentum of each individual wind fluid parcel is conserved, fixed by the rotation at its initial latitude at the wind base (i.e., at the subsonic stellar surface), and remaining in a fixed plane perpendicular to the angular momentum vector. The crucial criterion for the formation of a disk is that the near-star wind speed not be too much larger than the rotation speed. For lower rotation or faster winds, material may not be deflected into a disk, but still can form a more modest equatorial density enhancement, a "Wind Compressed Zone" (WCZ; Ignace, Cassinelli, and Bjorkman 1996).

To test this WCD paradigm, Owocki, Cranmer, and Blondin (1994) carried out 2-D hydrodynamical simulations of line-driven winds from rotating hot-stars. In keeping with the original WCD model, they assumed a purely radial driving force, though now computed dynamically using the finite-disk, spherical-star form of the usual CAK line-driving formalism (Friend & Abbott 1986; Pauldrach et al. 1986). The results, shown in figure 1a for their standard 'S-350' model (a B2 star with V_{rot} = 350 km/s), generally confirm the basic tenets of the WCD model, with certain detailed modifications (e.g., infall of inner disk material). The overall wind morphology consists of a relatively fast, low-density polar wind, plus a dense equatorial disk with slow outflow in its outer part. Figure 1a shows density contours for this standard WCD case, plotted vs. radius and colatitude, with superposed vectors representing the magnitude and sense of the latitudinal velocity component $v_\theta(r, \theta)$. The WCD, manifest here by the strong equatorial extension of higher density contours, is the direct result of the flow compression associated with the equatorward sense of the latitudinal velocity from both the northern and southern hemispheres.

3.2. INHIBITION OF WCD BY POLEWARD LINE-FORCE COMPONENT

For rotating stars, the line acceleration can generally also have *nonradial* components. Within the CAK formalism of a fixed ensemble of lines with

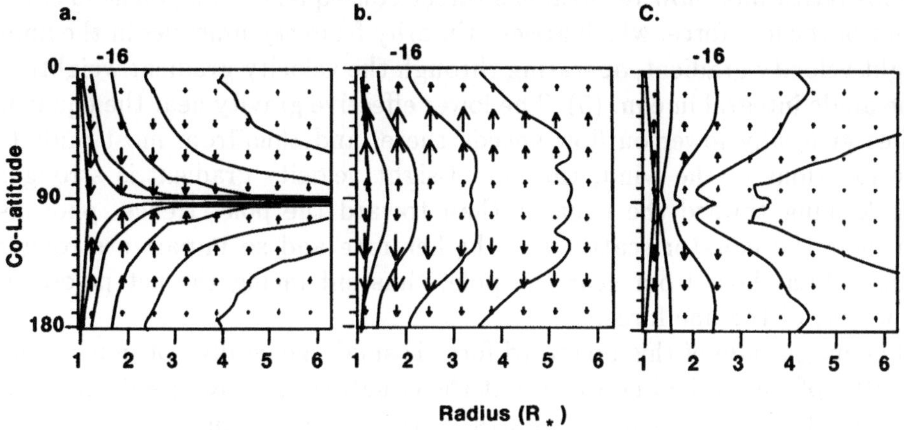

Figure 1. Contours of stellar wind density plotted vs. colatitude θ and radius r, spaced logarithmically with two contours per decade, with label denoting the contour for $\rho = 10^{-16} g/cm^3$. The superposed vectors represent the latitudinal velocity, with the maximum length corresponding to a magnitude of v_θ =100 km/s. The three panels show the cases (a) without nonradial forces or gravity darkening, (b) with nonradial forces but no gravity darkening, and (c) with both nonradial forces and gravity darkening.

a power-law number distribution in opacity, its vector form is given by integration over solid angle Ω_* of the stellar core intensity I (Cranmer and Owocki 1995),

$$g^{rad}(r) = \frac{K}{W^\delta \rho(r)^{\alpha-\delta} c} \int_{\Omega_*} d\Omega \, n \, I(n,r) \, \{n \cdot \nabla [n \cdot v(r)]\}^\alpha \, , \qquad (5)$$

where K is proportional to the usual CAK line-normalization constant k, and the exponent δ accounts for the ionization-state dependence on density ρ and dilution factor W (Abbott 1982). Note that the weighting for the force contribution of each ray along a direction n is proportional to the projected velocity gradient in that direction, $n \cdot \nabla [n \cdot v(r)]$.

Owocki, Cranmer, and Gayley (1996) carried out simulations of rotating winds including these nonradial line-force components. Figure 1b shows the corresponding wind density structure for the S-350 model, still assuming a uniformly bright stellar core. As predicted in the above 1-D analysis, the reduced gravity, enhanced mass loss, and lower flow speed near the equator yield a broad, moderate density enhancement in the equatorial wind. But there is *no wind compressed disk*! Indeed, the sense of the superposed vectors is now reversed, indicating that the latitudinal velocity is now *away from the equator.* As such, there is no longer any wind compression effect, and so the tendency to form a WCD is completely inhibited.

This latitudinal flow reversal is a direct consequence of a *poleward* component of the line-force, which arises primarily from asymmetries in the line-of-sight velocity gradient, operating through the velocity-gradient weighting of the angle integral in eqn. (5). The lower effective gravity near the equator implies generally lower outflow speeds there, and thus from most midlatitude locations in the wind, the line-of-sight velocity gradient is stronger when looking toward the equator than toward the pole. Hence, photons from near the equator scatter at a higher rate and so impart a stronger impulse than those from near the pole, thus enhancing the net poleward component of the line-force.

The magnitude of this poleward force is small, generally not much more than 10% of the radial line-force; but the equatorward flow speeds are similarly small, i.e. less than 100 km/s in the WCD model, or only a few percent of the maximum radial speed. Thus, while the radial line-force must be strong enough to overcome the stellar gravity to drive an outflow to terminal speeds of more than 1000 km/s, the poleward latitudinal line-force is unopposed by any other body force, and need only overcome inertial terms characterized by a modest, < 100 km/s equatorward drift. From this perspective, it thus seems clear that the derived nonradial forces should indeed be dynamically quite significant in redirecting the equatorward drift needed for a WCD. The next section presents a scaling analysis that quantifies these arguments.

3.3. STOPPING DISTANCE ANALYSIS FOR WCD INHIBITION

It is instructive to analyze this WCD inhibition effect in simple scaling-law terms. The general idea is to follow up on the above qualitative arguments that even a latitudinal force g_θ that is substantially smaller than than the radial force g_r should be able to stop the relatively slow v_θ associated with wind compression. Let us thus define a "stopping distance" s_o, over which the equatorward flow would be stopped if we suddenly applied the g_θ to a WCD model initially relaxed without nonradial forces. From simple mechanics, this is roughly given by

$$s_o \approx -\frac{v_\theta^2}{2g_\theta}. \tag{6}$$

Comparing this with $s_\theta \equiv r(\pi/2-\theta)$, the arc distance the material must move to get to the equator ($\theta = \pi/2$), we define a "stopping number",

$$\sigma \equiv s_\theta/s_o = -\frac{2g_\theta r(\pi/2-\theta)}{v_\theta^2}. \tag{7}$$

For σ order unity or larger, the nonradial forces should substantially alter or inhibit the wind compression. From the WCD models, we typically find an upper limit $v_\theta < V_{rot}/4$. To obtain an estimate for g_θ, we must examime the formulae in Cranmer and Owocki (1995). Ignoring, for simplicity, both gravity darkening and oblateness, the ratio of the latitudinal component of the line force to the point-star radial force can be written

$$\eta_\theta = \frac{r^2}{R_*^2(\partial v_r/\partial r)^\alpha} \int_{\mu_*}^1 d\mu' \int_0^{2\pi} d\phi' \left[n \cdot \nabla(n \cdot v)\right]^\alpha n_\theta, \tag{8}$$

where $n_\theta = \sqrt{1-\mu'^2}\cos\phi'$, and $\mu_* \equiv \sqrt{1 - R_*/r}$. The key is to find a simplification for the projected velocity gradient term, given explicitly by eqn. (41) of Cranmer and Owocki (1995). Following our argument that the main cause of the latitudinal force is the latitudinal gradient of the radial velocity, $\partial v_r/\partial\theta$, let us ignore all terms with v_θ or v_ϕ, keeping only the terms containing v_r. Assuming a characteristic value $\partial v_r/\partial r \approx v/r$, we can then write the projected gradient term as

$$\left[n \cdot \nabla(n \cdot v)\right]^\alpha = \left[\frac{\partial v_r}{\partial r} + \frac{\mu' n_\theta}{r}\frac{\partial v_r}{\partial \theta}\right]^\alpha \approx \left(\frac{\partial v_r}{\partial r}\right)^\alpha \left[1 + \alpha\mu' n_\theta \frac{\partial v_r/r\partial\theta}{\partial v_r/\partial r}\right], \tag{9}$$

where the latter approximation applies when the latitudinal gradient is small compared to the radial gradient. Applying this in eqn. (9), we find that the integrals can be carried out analytically, with the first term vanishing by the ϕ' integration of n_θ. After some manipulation (e.g. using the definition μ_*), we obtain

$$\frac{g_\theta}{g_r} \approx \eta_\theta \approx \frac{\alpha}{4}\frac{R_*^2}{r^2}\frac{\partial v_r/r\partial\theta}{\partial v_r/\partial r}, \tag{10}$$

where the initial approximation uses the fact that, in the wind region where $\partial v_r/\partial r \approx v/r$, the radial finite-disk correction factor is near unity, and so $\eta_\theta \approx g_\theta/g_r$. Then noting, from the equation of motion, that $g_r \approx v_r(\partial v_r/\partial r)/\alpha$ (i.e., the radial driving force is partitioned roughly in proportion α and $1-\alpha$ between inertia and gravity), we find

$$g_\theta \approx \frac{R_*^2}{4r^2}\frac{v_r}{r}\frac{\partial v_r}{\partial\theta} \tag{11}$$

Fig. 2b illustrates that this approximation is in good general agreement with numerical calculations of the line-force.

Finally, we need to parameterize the theta dependence of the radial velocity. For this, let us use the usual "effective escape speed" argument to write

$$v_r(r,\theta) \approx v_o(r)\sqrt{1 - \Omega\sin^2\theta}, \tag{12}$$

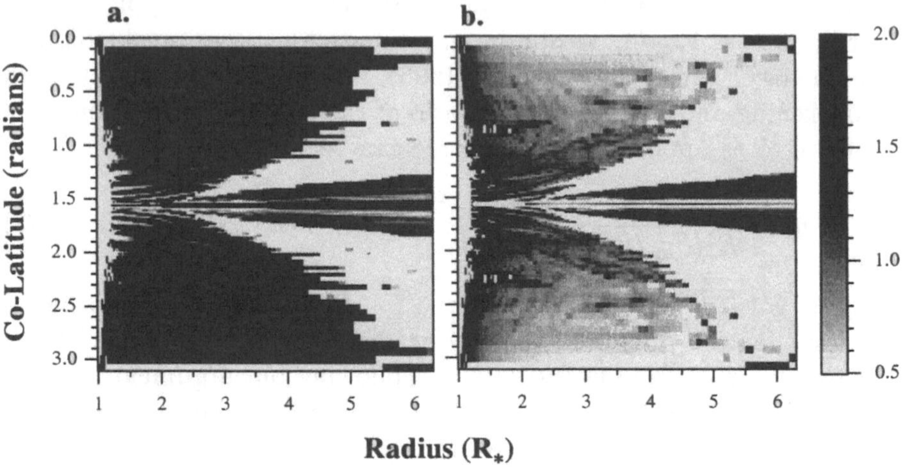

Figure 2. (a.) Grayscale plot of the stopping number, σ, in the S-350 WCD model, scaled over the range $0.5 < \sigma < 2$ for white to black. The dominance of black in almost the entire near-star region shows that the poleward component of the line force is sufficient to significantly alter or reverse the WCD tendency for material to deflect toward the equator. (b.) Grayscale plot of the ratio of the latitudinal force, g_θ, to the terms approximating it on the right-side of eqn. (11), also for the S-350 WCD model, with the same gray scale range as in (a). This shows that this approximation holds within a factor 2 throughout most of the region relevant to WCD inhibition.

where $v_o(r) \equiv v_r(r, \theta = 0)$ is the polar velocity law. Applying this in eqn. (11), we obtain

$$g_\theta \approx -\frac{R_*^2}{2r^2}\,\frac{v_o^2}{r}\Omega \sin\theta\cos\theta, \qquad (13)$$

which gives for the stopping number,

$$\sigma \approx \frac{v_o^2 R_*^2 \Omega \sin\theta\cos\theta\,(\pi/2 - \theta)}{v_\theta^2 r^2}. \qquad (14)$$

Evaluating this at a characteristic location $r = 2R_*$ and $\theta = \pi/4$, using $v_o(r = 2R_*) \approx v_{o,\infty}/2 \approx v_{esc}/2 = v_{crit}/\sqrt{2}$, and $v_\theta < v_{rot}/4$, we find $\sigma \gtrsim \pi/4$, indicating that this latitudinal force will have a substantial effect in inhibiting the wind compression.

Numerical calculations, e.g. for the S-350 model, confirm these estimates, yielding stopping numbers that range $1 < \sigma < 10$ over the entire wind compression region (fig. 2a). These same models also confirm the accuracy of the scaling estimate in eqn. (11), with surprisingly good agreement ($\lesssim$ factor 2) even in the areas where $\partial v_r / r\partial\theta$ is not necessarily small compared to $\partial v_r/\partial r$ (fig. 2b). The general conclusion is thus that the inibition

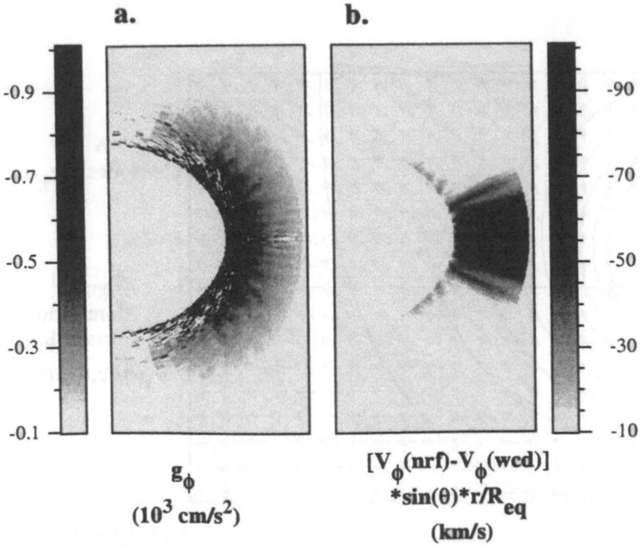

Figure 3. (a.) Grayscale plot of the azimuthal force, g_ϕ, in the S-350 WCD inibition model of fig. 1b. (b.) Grayscale plot of the azimuthal spindown of wind rotation as a result of the azimuthal line-force. This is given here by the decrease in azimuthal velocity in the WCD inhibition model (fig. 1b) from the initial WCD model (fig. 1a) without nonradial forces, scaled times $(\sin\theta)R_{eq}/r$, thus making this proportional to the loss of wind angular momentum parallel to the stellar rotation axis.

of WCDs is a quite robust effect, arising almost directly from the tendency of the radial outflow to be slower at lower latitudes.

3.4. SPINDOWN OF WIND ROTATION

Finally, although the 2-D models here have an assumed azimuthal symmetry, there is nonetheless also a nonzero *azimuthal* component of the line-force, which again results from asymmetries in the line-of-sight velocity gradient. Because wind rotation speed declines with increasing radius, the velocity gradient toward the receding stellar hemisphere is greater than that toward the approaching hemisphere. Through eqn. (5), this now implies a net line-force *against* the sense of rotation (Grinin 1978). Its peak magnitude is roughly comparable to that for the poleward line-force, but this is sufficient to cause a modest wind spindown, characterized by about a 25% decrease in the specific angular momentum of the equatorial wind outflow beyond a few tenths of a stellar radius from the surface. Fig. 3a shows grayscale plots of the azimuthal line-force in the S-350 WCD inhibition model (cf. fig. 1b), while fig. 3b shows the resulting azimuthal spindown of

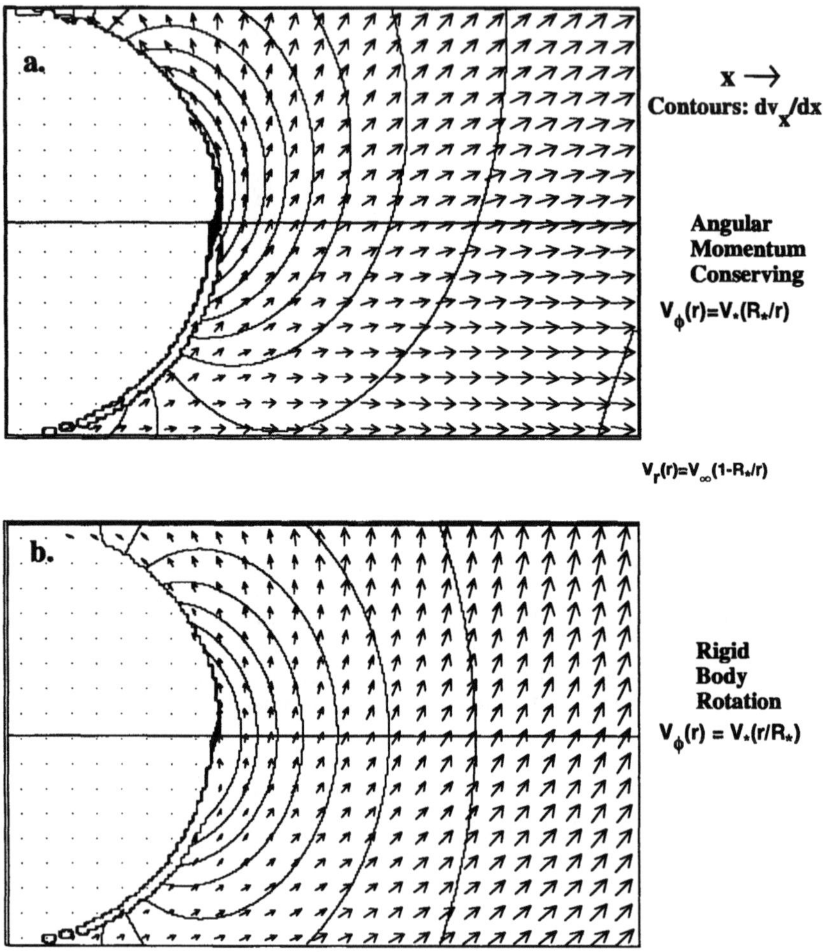

Figure 4. Contours of the projected velocity gradient $\partial v_x/\partial x$ in a direction x (oriented horizontally in the figure), plotted in the equatorial plane of a rotating wind with radial velocity $v_r(r) = v_\infty(1 - R_*/r)$ and azimuthal velocity $v_\phi(r)$ given either by: (a) angular momentum conservation, $v_\phi(r) = v_* R_*/r$; or (b) rigid body rotation, $v_\phi(r) = v_* r/R_*$. The superposed vectors indicate the sense and magnitude of the total velocity field.

the wind in this model. Note that the equatorial regions of the wind have lost more than 100 km/s of their 350 km/s surface rotation speed.

Fig. 4 provides a graphical illustration of the origin of the azimuthal asymmetry in the line resonance that gives rise to a net azimuthal line-force. Consider the flow in the equatorial plane of a simple wind model with a radial velocity given by the ('beta=1') law $v_r(r) = v_\infty(1 - R_*/r)$, and the azimuthal velocity set according to angular momentum conservation, $v_\phi(r) = v_* R_*/r$ (fig. 4a), or to rigid-body rotation, $v_\phi(r) = v_* r/R_*$ (fig. 4b). The contours are of the projected velocity gradient, $\partial v_x/\partial x$, in a given direction x, oriented horizontally on the figure. Because of the azimuthal symmetry of the radiation field, the azimuthal line-force is proportional to the integral,

$$g_\phi \sim \left[\int_{-R_*}^{R_*} dy\, y\, \frac{\partial v_x}{\partial x} \right]_r , \tag{15}$$

taken along an arc of constant radius r, over the range $-R_* < y < +R_*$, where y is the vertical coordinate in the figure. The key point is that in fig. 4a, the projected velocity gradient is biased toward the upper part of the plane, indicating the net integral will give a net line-force from radiation originating in the upper, i.e. receding, portion of the stellar surface. By contrast, in the case of rigid rotation shown in fig. 4b, the projected velocity gradient is *symmetric* between the upper and lower half of the plane, and so the azimuthal line-force vanishes in such a rigid rotation wind. Indeed, in the somewhat unlikely case that the azimuthal velocity should have a faster than rigid-body outward increase, the bias would be toward the lower half-plane, leading to a net *spin-up* line-force!

In the other extreme, if the azimuthal velocity falls off faster than for angular momentum conservation, the spin-down force would be stronger. In principle, this raises the possibility of a "runaway wind spindown" solution, in which the azimuthal line-torque leads to a sufficiently steep radial fall-off of azimuthal velocity that the line-torque becomes ever stronger. In practice, we find that overall strength of the line-torque is just a little (i.e. $<$ factor 2) too small to give such runaway solutions, except in rather artificial circumstances, i.e. very low value of α ($\ll 0.5$) or a strong stellar limb brightnening. Though the possibility of a azimuthal line-force was first recognized long ago (Grinin 1978), these latest studies suggest that it can come very close to being a dominant effect. We intend to discuss these analyses further in a future paper.

4. Effect of Gravity Darkening

4.1. UNIFORM WIND DRIVING PARAMETERS

Fig. 1c shows results for the corresponding S-350 model with both nonradial forces and a gravity-darkened surface flux $F(\theta) \sim g_{eff}(\theta)$ (von Zeipel 1924; see also Cranmer & Owocki 1995). In this case, not only is there no disk, but the overall density in the equatorial regions is actually *reduced* relative to that at higher latitudes. This picture is in marked contrast with previous analyses that envisioned an enhanced equatorial mass loss [e.g., eqn. (3); Friend and Abbott 1986], but it agrees well with the predictions of the gravity-darkened mass-flux scaling (4). Despite the reduced gravity near the equator, the wind mass flux there is now lower, owing to the reduced radiative flux associated with gravity darkening. The superposed vectors further show that the latitudinal velocity is again away from the equator, though with a somewhat lower magnitude than in Figure 1b, owing to the reduced poleward force associated with the reduced radiative flux from the equator.

4.2. EQUATORIAL BI-STABILITY ZONE

The above calculations have assumed fixed values of the CAK parameters (α, k, and δ), but in general these can be expected to vary with variations in, e.g., effective temperature and density. A particularly important example of this is the Lamers and Pauldrach (1991) *Bi-Stability* model for equatorial enhanced winds in B[e] stars. At effective temperatures near 20,000 K, the increased recombination of hydrogen tends to make a wind become optically thick in the Lyman continuum, thus dramatically altering the wind ionization/excitation balance, and so leading to a marked shift of line-driving parameters. The effect can be most easily described in terms of a decreasing CAK exponent α, with a relatively constant line normalization parameter $\bar{Q} \approx 10^3$, related to the usual CAK k constant by $k = \bar{Q}^{1-\alpha}(v_{th}/c)^\alpha/(1-\alpha)$ (Gayley 1995). Within CAK theory, the mass flux varies as $\dot{m} \sim \bar{Q}^{-1+1/\alpha}$, and so adopting this into the scaling formulae in §2, we find for the latitudinal variation of density $\rho \sim \dot{m}/v$,

$$\frac{\rho(\theta)}{\rho_o} = \sqrt{1-\Omega\sin^2\theta}\,(\Gamma_e\bar{Q})^{1/\alpha-1/\alpha_o}, \tag{16}$$

where $\Gamma_e = \kappa_e F/g_{eff}c$ is the surface Eddington factor, which is independent of latitude in the standard von Zeipel (1924) scaling $F \sim g_{eff}$ for gravity darkening. The exponent α is now assumed to vary in latitude, e.g. from a typical O-star value $\alpha_o \approx 2/3$ at the relatively high-temperature pole, to a lower, B-star value $\alpha \approx 1/2$ at lower latitudes, where the rotation brings the effective temperature near or below the critical bi-stability temperature of $\sim$ 20,000 K. The expression here includes the reduced driving from the lower equatorial brightness, an effect that actually was overlooked in the original Lamers and Pauldrach (1991) analysis, cf. their eqn. (4) and our eqns. (3) and (4). For stars with Eddington factors Γ_e not much less than unity, the large intrinsic value of $\bar{Q}$ implies that this shift in α can cause a strong increase in mass flux, as signified by the second term in eqn. (16). For example, for the case $\Gamma_e = 0.5$, $\bar{Q} = 10^3$, $\alpha_o = 2/3$, and $\alpha = 1/2$, this bistability density jump represents a quite large factor, $\sim \sqrt{500} \approx 22$. In principle, this could occur quite abruptly near and below the latitude of the critical temperature, and thus overwhelm the more gradual tendency for the mass flux to decline with the decreasing radiative flux near the equator. As such, this model remains a viable possibility for explaining moderate increases in equatorial density, such as inferred for B[e] stars.

5. Concluding Remarks

The above results on nonradial forces and gravity darkening are both a surprise, in that they essentially reverse previous theoretical expectations, and a puzzle, in that they apparently contradict observational evidence for enhanced densities in the equatorial regions around rapidly rotating hot-stars. As regards gravity darkening, if convection were to dominate energy transport in the envelope of rapidly rotating stars, then the equatorial darkening could be substantially reduced, or even eliminated (Zhou and Leung 1990). Given the formidable difficulties of multidimensional modelling of convection in rapidly rotating stellar envelopes, this emphasizes the importance of determining reliable observational diagnostics of gravity darkening in such stars (e.g., Howarth and Reid 1993; Reid et al. 1993).

As regards WCD inhibition by nonradial line-forces, we emphasize the following point:

> The WCD inhibition described above is specific to *line-driven* winds, and does *not* imply a failure or weakness of the general WCD paradigm as potentially applied to outflows driven by any other mechanism.

The great success and intuitive appeal of the WCD paradigm stems from the simplicity of its key idea, that material slowly driven radially outward from a star should, by virtue of angular momentum conservation, remain in an orbital plane that intersects the equator. In the context of line-driven flows, however, this key assumption of strictly radial driving represents a serious oversimplification, because of the nonvanishing poleward component of the line-force, which can retard and indeed reverse the equatorward flow essential for a WCD. Moreover, within the standard CAK/Sobolev formalism, this WCD inhibition seems quite robust, arising directly from the tendencies for the line-force to scale with the line-of-sight velocity gradient, and for the wind from a rotating star to be slowest at lower, near-equatorial latitudes. Insofar as winds from O and B stars have traditionally, and quite successfully, been modelled using the CAK, line-driving formalism, this clearly represents a serious challenge for applying the WCD paradigm toward, e.g. Be stars, B[e] stars, and selected O-stars (e.g., HD 93521: Howarth and Reid 1993; Bjorkman et al. 1994) inferred to have enhanced equatorial densities and/or disks. As such, perhaps the findings here can be interpreted as indicating that other mechanisms, e.g. decretion, magnetic loops, or bistability zones, should be examined as ways to produce the disks or equatorial density enhancements inferred in such stars.

However, there are also still questions regarding the applicability of such idealized, CAK line-driven models, particularly for B-star winds. (See, e.g., Cassinelli and Ignace 1997, as well as comments and contributions by J. Cassinelli and J. Bjorkman in these proceedings.) One particularly serious general question (emphasized to us by J. Bjorkman) regards the neglect so far of the strong line-driven flow instabiilty (see, e.g., Owocki 1994 and references therein), which ultimately should break the wind up into multiple dense blobs that might not receive the same poleward driving force. This question needs to be addressed in future work, but it represents a formidable computational challenge to develop tractable methods for evaluating the required *nonlocal* escape integral forms for the line-force (Owocki and Puls 1996) within a multi-dimensional dynamical simulation. Given these remaining questions, it may not yet be appropriate to strictly rule out the WCD model, even for line-driven models of B[e] stars. But though not dead, the line-driven WCD does seem seriously crippled, or at least 'dynamically challenged'.

Whatever the ultimate resolution to these puzzles, we hope that the results here will spur a reexamination and questioning that ultimately leads to a deeper understanding of the nature of hot-star mass loss. back. It is remarkable that, more than two decades after introduction of the line-driving mechanism in landmark papers by Lucy and Solomon (1970) and CAK, we are still learning fundamentally new aspects of that intricate process.

Acknowledgements

This work was supported in part by NASA grant NAG5-3530. We thank J. Bjorkman, J. Cassinelli, N. Langer, and J. Puls for useful discussions.

References

Abbott, D.C. 1982, ApJ, 259, 282
Bjorkman, J. E., & Cassinelli, J. P. 1993, ApJ, 409, 429
Bjorkman, J. E., Ignace, R., Tripp, T., & Cassinelli, J. P. 1994, ApJ, 435, 416
Castor, J. I., Abbott, D. C., & Klein, R. I. 1975, ApJ, 195, 157 (CAK)
Cassinelli, J. P. & Ignace 1997, in *Luminous Blue Variables: Massive Stars in Transition*, ASP Conf. Series 120, A. Nota & H.J.G.L.M. Lamers, eds., (PASP: San Francisco)
Cranmer, S. R., & Owocki, S.P. 1995, ApJ, 440, 308
Friend, D. B., & Abbott, D. C. 1986, ApJ, 311, 701
Gayley, K. G. 1995, ApJ, 454, 410
Grinin, A. 1978, Sov. Astr., 14, 113
Howarth, I. D. & Reid, A. H. N. 1993, A&A, 279, 148

Ignace, R., Cassinelli, J. P., and Bjorkman, J. E. 1996, ApJ, 459, 671
Kippenhahn, R. & Weigert, A., 1990, *Stellar Structure*, (Springer: Berlin), p. 436
Lamers, H. J. G. L. M., and Pauldrach, A. W. 1991, A&A, 244, L5
Lucy, L. B. & Solomon 1970, ApJ, 159, 879
Owocki, S. P. 1994, Ap. Sp.Sci., 221, 3.
Owocki, S. P., Cranmer, S. R., & Blondin, J. M. 1994, ApJ, 424, 887
Owocki, S. P., Cranmer, S. R., Gayley, K. G. 1996, ApJ, 472, L115.
Owocki, S. P. & Puls, J. 1996, ApJ, 462, 894
Pauldrach, A.W.A., Puls, J. & Kudritzki, R.-P. 1986, A&A, 164, 86
Reid, A. H. N. et al. 1993, ApJ, 417, 320
von Zeipel, H. 1924, MNRAS, 84, 665
Zhou, D. & Leung, K. 1990, ApJ, 355, 271

Discussion

Cassinelli: Your conclusion regarding wind compression came across very strongly against the idea. However the basic idea should work for the red supergiants and other non CAK winds. Another important difference is that the basic wind compression models are kinematic models, while yours is dynamic and so it requires that the CAK approximation you use are valid. For example your CAK based models have a very quickly accelerating outflow ($\beta \approx 0.8$) and as I mentioned in my talk wind compression is less likely than is the case for more slowly accelerating winds. If the wind velocity law is slower, the star would not need to have the large rotation speeds that you use. If the stars are rotating at say 30 % break-up, the distortion, equatorial darkening and Sobolev effects are greatly reduced. Finally in your models you do not account for variations of the CAK, κ, α, δ parameters as a function of latitude or radius. If κ decreases with radius, would not the velocity law be slower and the formation of a WCD more likely?
Owocki: As I tried to emphasize in my talk, these results do NOT imply that WCDs are in any way inhibited for non-CAK winds, like in AGB's. Perhaps there are difficulties in applying CAK theory to B[e] or other hot stars, e.g. fitting observations that imply a high β velocity law. But CAK is the standard theory for winds of these hot stars and the results have seen to follow directly from basic physics of this mechanism. I have begun to consider effects of varying CAK κ, α with latitude, as in the discussion of the Lamers and Pauldrach bi-stability model. Having κ decline with radius may indeed provide a way to get a slower velocity law, for which the standard WCD theory with only radial line forces would predict stronger disk. But this should still not affect the non radial line force that arises from the slower wind in equator.vs.pole. However I will explore this case in future models.
Bjorkman: As you have demonstrated, small forces of the order of a few percent of gravity are important for determining the flow structure. Al-

though you have considered many possible effects, there are still others that may change the line driving force. For example, clumping in the wind (with a small velocity dispersion within the clump) would make the clump size the length scale for calculating the line optical depths instead of the Sobolev length. In this case the non-radial forces are determined by the shape of the clump instead of the velocity gradient in a smooth flow. Another possible effect that you have not included is photospheric line structure. Owing to the stellar rotation, the effect of a large number of photospheric lines will be to produce a randomly varying non-radial component to the force. Is it possible that such a randomly varying force could produce clumping in the wind, which in turn would change the nature of the driving forces on the clumps as I discussed above?

Owocki: Such complications certainly need to be investigated. But I would emphasize that the non-radial effects are only small compared to gravity; they are of the same order as the anticipated WCD flow. That is, in this sense, the original WCD model also depends on small terms. Though I can never say that new effects would not modify things, I believe that the non-radial effects are quite robust in that they depend on very basic characteristics of a CAK wind, namely the lower expected speed in the equator.vs.pole and the line force scaling with the line-of-sight velocity gradient to the radiation source. But certainly issues, such as clumping need to be investigated, especially since such clumping is expected from small scale line driven flow instabilities that are NOT considered in the Sobolev-theory treatment of the line forces used here.

Araujo: From your computations the bi-stability mechanism (which means changing the relative contribution of the driving lines) is essential to obtain a density enhancement in the equatorial region; can I summarize like that?

Owocki: Yes. WCDs appear to be inhibited in line driven flows, but bi-stability is still possible for B[e] stars, though somewhat reduced by gravity darkening. It thus remains a possible explanation, along with other mechanisms (e.g. decretion disks spun up by magnetic fields or non-radial pulsations), mentioned in Cassinelli's review talk.

MASS LOSS RATES FROM B[e] STARS

J.A. DE FREITAS PACHECO
Observatoire de la Côte d'Azur, BP 4229 06304 Nice Cedex 4, France

Abstract.
Mass loss rates for a sample of eight blue supergiants are calculated using the hydrogen lines Hα, Hβ and Pδ. B[e] stars have a $\dot{M}$ vs L relationship flatter than that of OB stars, and mass loss rates comparable to those observed in WR stars.

1. Introduction

B[e] stars are massive objects (M > 20 $M_\odot$), which have already evolved off the main sequence, and most of them are in the He-burning phase. These stars have quite massive winds responsible for a substantial loss of their initial mass. In those objects the "effective" photosphere is in the wind itself, and therefore out of hydrostatic equilibrium. Like the wind of WR stars, multiple scattering of photons is necessary to explain the total rate of momentum carried out by the flow (Springmann 1994).

These stars as a "class" share the following common properties: a) no photospheric features are (usually) seen in the optical spectrum, masked by the dense wind; b) the optical spectrum is dominated by emission lines of HI and FeII, often displaying P Cyg profiles. HeI lines are present only in early B types; c) the UV spectrum is dominated by an "iron curtain" (abs/em), and some other low ionization species like MgII, TiII, MnII may also be present (de Freitas Pacheco et al. 1982, 1985; Shore et al. 1990); d) most of these stars are IRAS sources; e) in the near infrared, besides HI lines, the most prominent emission features are those of FeIIλ999.7 nm, OIλ844.6 nm and the CaII red triplet at λ850,854,867 nm (Lopes et al. 1992); f) variability in the continuum and in the line profiles with different timescales; g) significant linear polarisation, often variable.

A. M. Hubert and C. Jaschek (eds.), B[e] Stars, 221–225.

Some of these objects have probably ejected a substantial amount of matter 10^4 years ago, which are now forming a shell nebula around the star. These nebulae have diameters typically in the range 1-2 pc and masses of 2-4 $M_{\odot}$. The high N/O ratio observed in these surrounding nebulae suggests that such a material was nuclearly processed by the progenitor. In fact, such high N/O ratios are expected at the LBV phase (Maeder & Meynet 1988), indicating that those nebulae may be an important tool for our understanding of the chemical changes suffered by the progenitor. The origin of this low velocity material (40-70 km/s) is not yet well established. If the star has returned back from a red supergiant phase, then the gas could be the remnant of the wind existing at that epoch. However this might only occur for stars with initial masses less than 40 $M_{\odot}$, which is not the case of AG Car (de Freitas Pacheco et al. 1992; Viotti et al. 1993). Another possibility is mass ejection induced by pulsations during the unstable LBV phase. Besides AG Car, stars associated to shell nebula are HD 87643 (reflection neb.), He 3-519 (em. neb.), He 3-1300 (em/reflec neb.), HR Car (em. neb.), WRA 751 (em. neb.), η Car (em/reflec neb.) and P Cyg (em. neb.).

2. Mass loss rates

The wind in B[e] stars plays a major role in the evolution and in the observed properties of these objects. Accurate mass loss determinations are necessary to better understand the evolution of these stars and the driving wind mechanism.

TABLE 1. Physical Properties

Object	$\log(L/L_{\odot})$	$R/R_{\odot}$	V_{∞}(km/s)	$M/M_{\odot}$
HD 87643	5.46	68	1000	34
HD 89249	5.22	40	650	25
HR Car	5.62	117	1000	32
GG Car	5.18	30	670	25
HD316285	5.98	48	1300	60
HD327083	6.26	260	700	68
CPD-52^{0}9242	5.90	113	800	48
P Cyg	5.97	76	400	60

In this contribution, mass loss rates are derived for a sample of eight blue supergiants, using HI emission lines (Hα, Hβ and Pδ). The main physical parameters of these stars are given in table 1. Initial masses were estimated from the HR diagram and tracks by Maeder & Meynet (1988). The observed luminosities (in erg/s) are given in table 2 and correspond to data obtained at Pico dos Dias (Brazil).

The wind luminosity in a given transition is given by

$$L_{ij} = 4\pi \int_{env} N_j A_{ji} h\nu_{ij} \beta_{ij} (1 - \beta^*_{ij}) r^2 dr \tag{1}$$

The level population density N_i was calculated from the statistical equilibrium equations (eight level atom), and the total matter density ρ satisfy the conservation equation

$$\dot{M} = 4\pi \rho V(r) r^2 \tag{2}$$

The photon escape probability β_{ij} and the probability of striking the core β^*_{ij} were calculated using the Sobolev theory. The adopted velocity profile was

$$V^2(r) = V_0^2 + (V_\infty^2 - V_0^2)(1 - \frac{R}{r})^\gamma \tag{3}$$

In the above equation, the so called CAK velocity profile is recovered when $\gamma = 1$ (most of our cases).

TABLE 2. Line luminosities & mass loss rates

Object	log(Hα)	log(Hβ)	log(Pδ)	log($\dot{M}$)
HD 87643	37.11	36.39	35.17	-4.30±0.06
HD 89249	35.96	35.70	-	-5.03±0.04
HR Car	36.33	35.92	35.59	-4.32±0.22
GG Car	36.29	35.82	33.48	-4.88±0.18
HD316285	37.12	36.58	35.33	-4.12±0.06
HD327083	37.09	36.34	35.40	-4.11±0.07
CPD-52^{0}9242	36.70	36.48	35.30	-4.27±0.03
P Cyg	36.56	-	34.90	-4.39±0.20

These equations were solved numerically (an abundance ratio $\frac{N(He)}{N(H)} =$ 0.15 was adopted), searching for mass loss rates (and γ's) which give the best representation of the observed luminosities. The last column of table 2 shows the resulting mass loss rates and errors estimated from the fit quality.

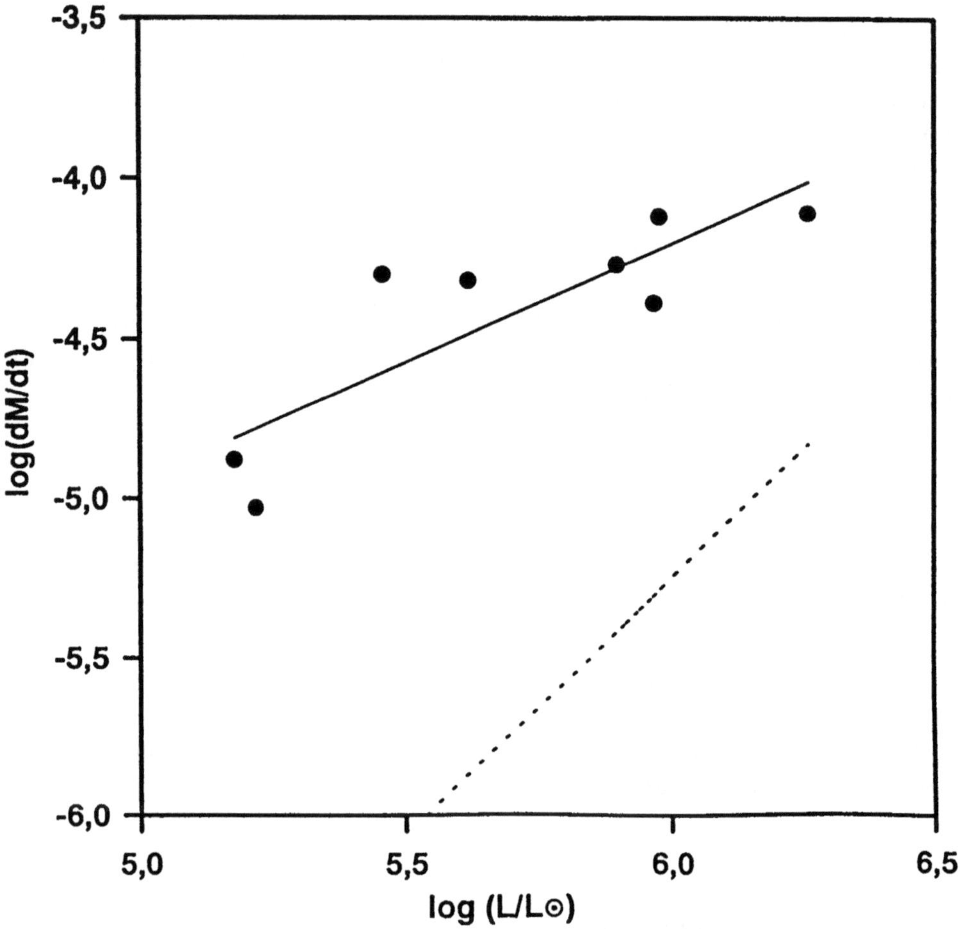

Figure 1. Mass loss in solar masses per year vs. bolometric luminosity. The traced curve below corresponds to the relation observed for OB stars

In figure 1, the derived mass loss rates are plotted against the bolometric luminosities. The solid curve is the best fit to the data, given by the equation

$$log(\dot{M}) = -8.64 + 0.74 log(\frac{L}{L_{\odot}}) \pm 0.20 \qquad (4)$$

whereas the traced curve represents the same relation for OB stars.

3. Conclusions

From inspection of figure 1, it is clear that B[e]'s obey a relation $\dot{M}$ - L different than that followed by OB stars. Firstly, the dependence of $\dot{M}$ on luminosity is flatter and secondly, for a given L, B[e]'s have a higher mass loss rate, comparable to values observed for WR stars. Therefore these stars, like WR's, need a more efficient mechanism of momentum (and/or energy) transfer from the radiative field to the gas in order to drive the wind.

The efficiency ε_W of energy conversion required to drive the wind is $\varepsilon_W \approx \frac{\dot{M}V_\infty^2}{2L_\infty}$ and, from the present data is in the range 0.002 - 0.02. The efficiency ε_W correlates with the characteristic wind density ($\varepsilon_W \propto \rho^{0.8}$) supporting results from numerical simulations of multiple scatering of photons as the main acceleration mechanism of the wind.

References

de Freitas Pacheco J.A., Gilra D., Pottasch S.R., 1982, A&A 108, 111
de Freitas Pacheco J.A., Lopes D.F., Landaberry S.J.C., Selvelli P.L., 1985, A&A 152, 101
de Freitas Pacheco, J.A., Damineli A., Costa R.D.D., Viotti R., 1992, A&A 266, 360
Lopes D.F., Damineli A., de Freitas Pacheco J.A., 1992, A&A 261, 482
Maeder A., Meynet G., 1988, A&AS 76, 411
Shore S.N., Brown D.N., Bopp B.W., Robinson C.R., Sanduleak N., Feldman P.D.,1990, ApJS 73, 461
Springmann U., 1994, A&A 289, 505
Viotti R., Polcaro V.F., Rossi C., 1993, A&A 276, 432

INFRARED EMISSION FROM DUST AROUND B[e] STARS

IVEZIĆ ŽELJKO
Princeton University, Department of Astrophysical Sciences, Princeton, NJ 08544, USA

MIROSHNICHENKO ANATOLY S.
Central Astronomical Observatory of the Russian Academy of Sciences at Pulkovo, Saint-Petersburg, 196140, Russia

AND

ELITZUR MOSHE
Department of Physics and Astronomy, University of Kentucky, Lexington, KY 40506-0055, USA

Many B[e] stars show IR excess indicating the presence of circumstellar dust. However, HD 45677 is the only such star with a detailed model of its spectral energy distribution (SED), computed by Sorrell (1989). Here we present such modeling in the wavelength range 0.3 – 100 μm for a sample of 13 B[e] stars.

The model SEDs are computed with code DUSTY which provides an exact solution of the spherically symmetric dust radiative transfer problem (Ivezić, Nenkova, & Elitzur 1997). Stellar radiation is described by Kurucz (1979) models for a given spectral type, dust density distribution by a power law r^{-p1}. When a satisfactory fit cannot be obtained with a single power law, we assume a second power-law, r^{-p2}, between $r_p = Y_p r_{in}$, where r_{in} is the inner radius determined by the dust sublimation temperature T_{sub}, and the outer radius $r_{out} = Y_{out} r_{in}$. We employ dust optical properties for the standard interstellar dust (Draine & Lee 1984). The envelope's overall visual optical depth τ_V, interstellar extinction A_V, T_{sub}, $p1$, and Y_{out} are the free model parameters (and also Y_p and $p2$ when necessary).

Examples of the best fits are shown in Fig. 1, and the corresponding model parameters in Table 1. In most cases the model SEDs agree well with the observations. The dust density distributions can be roughly described as $r^{-3/2}$ in the inner shell, and by somewhat flatter fall-off in the outer regions.

A. M. Hubert and C. Jaschek (eds.), B[e] Stars, 227–228.

TABLE 1. Best-fit model parameters

Name	IRAS	A_V	τ_V	T_*	T_{sub}	p1	Y_p	p2	Yout
AS 78	03549+5602	2.1	0.4	16000	1500	1.7			700
HD 45677	06259−1301	1.3	0.2	25000	800	1.7			1000
HD 50138	06491−0654	0.2	0.6	10000	1500	1.5	50	0.5	500
3 Pup	07418−2850	0.0	0.4	10000	1500	1.5			2000
Ve 2–27	08502−4606	4.3	0.2	25000	800	1.3	500	0.5	5000
HD 87643	10028−5825	0.0	4.5	10000	1500	1.0			5000
V921 Sco	16555−4237	0.0	30.0	10000	1500	1.5	500	0.0	5000
MWC 922	18184−1302	0.0	10.0	10000	600	1.5			5000
MWC 300	18267−0606	4.5	0.1	25000	800	1.7	1000	0.0	10000
MWC 939	18313−1738	2.6	0.2	25000	1100	1.5	50	2.0	5000
V1972 Cyg	20212+3920	4.6	0.14	30000	1500	1.5			5000
V1478 Cyg	20310+4029	9.6	0.01	30000	1400	1.0			5000
MWC 657	22407+6008	4.9	0.05	20000	1500	1.7			700

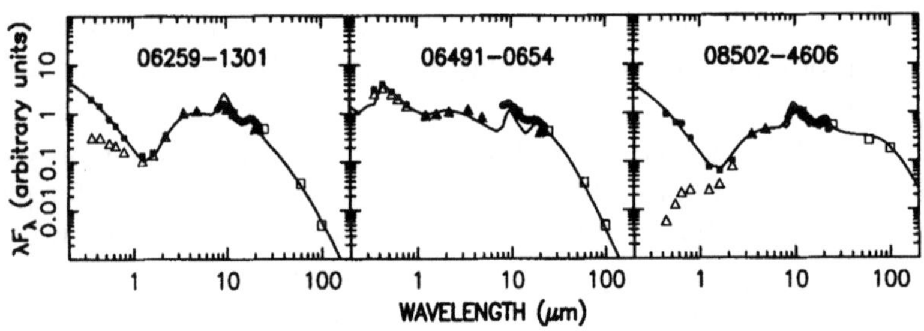

Figure 1. Observations and model results. IRAS PSC data are denoted by circles and IRAS LRS by open squares. Triangles denote all other data without any correction for interstellar extinction, filled squares are the same data corrected for extinction with the listed A_V. The lines are the results of the modelling described in the text.

We conclude that in addition to the heterogeneous spectroscopic behavior of B[e] stars, their SEDs also differ significantly within the group, possibly reflecting different evolutionary status of individual stars.

References

Draine, B.T., Lee, H.M, 1984, *ApJ*, **285**, 89
Ivezić, Ž., Nenkova, M., & Elitzur, M.: 1997, *Univ. of Kentucky, Internal Report*
Kurucz, R.L. 1979, *ApJS*, **40**, 1
Sorrell, W.H. 1989. *MNRAS*, **241**, 89

AXISYMMETRIC MODEL ATMOSPHERES OF B[e] STARS

T. LANZ
Sterrenkundig Instituut, Utrecht University
P.O. Box 80000, NL-3508 TA Utrecht, The Netherlands

Abstract. Initial results of an investigation of B[e] stars infrared spectrum are presented. New axisymmetric model atmospheres have been developed to analyze ISO observations of B[e] supergiants. The model atmospheres include a spherically symmetric wind and a dense, flat, outflowing equatorial disk that are described by parametric density and velocity laws. The radiative transfer is then solved exactly, and the sensitivity of infrared lines to various geometries is explored.

1. Introduction

A schematic two-component model was proposed by Zickgraf et al. (1985) to explain the hybrid spectrum of LMC B[e] supergiants. Broad, high-excitation UV features, like C IV, Si IV, and N V resonance lines, are formed in a fast, hot polar wind, while narrow permitted and forbidden low-excitation emission lines are formed in dense, equatorial disk. Strong Balmer emission lines are a most typical characteristics of B[e] stars. Both components, the fast wind and the dense inner disk, contribute to the emission. Modeling these lines requires thus to incorporate the two components simultaneously and consistently in the model. B[e] stars are also characterized by a strong mid-IR excess, due to the presence of hot dust in the disk. Dust and molecules can form in the outer disk at distances, $R \approx 500 - 10^3\, R_\star$, thanks to the shielding of stellar radiation provided by the disk.

This general paradigm explains the spectral characteristics of B[e] stars, and describes the physical conditions in the circumstellar environment of these stars. Similar environments around stars in different evolutionary stages (PMS, supergiants, proto-PN) produce similar spectra, and a spectral analysis will therefore reveal more on the wind and equatorial disk

A. M. Hubert and C. Jaschek (eds.), B[e] Stars, 229–233.

than on the central star, especially since no photospheric lines are observed (Zickgraf 1998).

Bjorkman & Cassinelli (1993) and Ignace et al. (1996) attribute the formation of an equatorial disk to a wind in fast rotation. Material flowing along stream lines will accumulate at the equator providing that the wind and rotation velocities are of similar order. High density contrast between the polar and equatorial regions in B[e] supergiants are predicted already for moderate rotation rates ($\Omega \approx 0.3 - 0.4\,\Omega_{crit}$) if the wind is slowly accelerating ($\beta = 3$). However, detailed 2-D dynamical simulations (Owocki et al. 1998) suggest that it might not be the case due to a non-radial component of the line force. Lamers & Pauldrach (1991) proposed the so-called "bi-stability" mechanism as an alternative way to form an equatorial disk. Basically, lower radiation temperature in the equatorial regions will trigger enhanced mass-loss at the equator due to the change in the line force.

Starting from a schematic picture, the following step consists in developing a quantitative, empirical model aiming to discriminate between the different mechanisms put forward to explain the B[e] phenomenon. Models characterized by an axial symmetry with a wind and a disk have already been developed for classical Be stars (Poeckert & Marlborough 1978, Waters 1986, Stee et al. 1995) and for cataclysmic variables (e.g. Shlosman et al. 1996, Knigge et al. 1995). My approach to this problem consists in solving the radiative transfer exactly in a prescribed geometry that is described by simple, parametric laws for the density and velocities. I derive then spectral diagnoses of the wind and disk structure by matching the observed profiles of IR hydrogen recombination lines. Such an quantitative empirical model may be compared later to theoretical predictions.

I present here briefly the parametric model, a short discussion of Brα sensitivity to the model parameters, and some initial results on B[e] disks based on ISO-SWS spectra. A complete description can be found in a paper submitted to A&A.

2. A parametric model

Assuming an axisymmetric geometry around the rotation axis, all quantities can be described as functions of the distance to the star's center (r) and of the polar angle (θ). I assume a density law of the following form:

$$\rho(r,\theta) = \rho_0\,(\frac{r_0}{r})^n\,(1 + a\,\sin^m(\theta))$$

where r_0 is the star radius, ρ_0 the polar density at the star surface, and n an exponent for the radial dependency (typically n is about 2.5). The density contrast between the pole and the equator is set by a, while m is related to a typical opening angle of the disk. The exponent m becomes

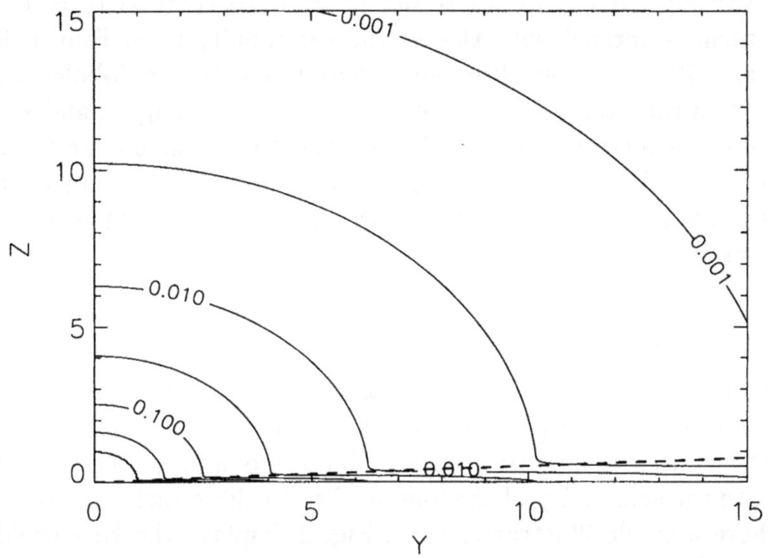

Figure 1. Iso-density curves for a model with $\rho_{eq}/\rho_p = 10$ and $\theta_d = 3^o$.

rapidly quite large for small opening angles. Defining the disk boundary as where the density is 20% larger than the polar density, $m = 2852$, for an opening angle of 3° and a density contrast of 10. Iso-density curves for this example are shown in Fig. 1.

The expansion velocity field is described by a β-type law, though the terminal velocity v_∞ and the acceleration parameter β can vary with the polar angle to allow for fast winds in the polar region and slow winds in the disk. In addition, rotation is included assuming conservation of the angular momentum. The expansion and rotation velocities are expressed in the orbital plane and then transformed back to the stellar coordinates giving a 3-D velocity field (see Bjorkman & Cassinelli 1993).

A grey temperature structure is assumed, limited to some minimum value, T_w, in the wind and, T_d, in the disk. Ionization structure and non-LTE b-factors of hydrogen levels are also dependent on the polar angle. They are derived from a series of non-LTE spherically-symmetric models pieced together. The non-LTE source function combines a scattering term and a thermal term:

$$S = \frac{(1-\varepsilon) < \beta I > + \varepsilon\, S_t}{(1-\varepsilon) < \beta > \varepsilon}$$

The thermal part is the thermal source function (and not the Planck function like in the two-level atom); thus, ε is more like a parameter that can be chosen varying from pure scattering ($\varepsilon = 0$) to thermal ($\varepsilon = 1$). The scattering term is derived with the escape probability formalism. β is the escape probability is a given direction calculated with the Sobolev optical depth, $< \beta >$ is the average escape probability over all angles, and $< \beta I >$ is the so-called penetration probability averaged over the stellar disk. The contribution of disk scattering is neglected. The radiative transfer along about 2500 rays is then solved exactly through the line profile, using the discontinuous finite element scheme (Castor et al. 1992).

3. Initial results

The parametric model includes many parameters, describing the central star, the wind and disk structures, the source function, and the geometrical aspect of the system (inclination angle). Therefore, a first necessary step is to investigate the sensitivity of the line profiles to these various parameters. I present here a single illustrative case. Fig. 2 displays the Brα profile for 4 different disk opening angles, assuming that all other parameters are identical. In particular, a density contrast of 10 is assumed, the wind is fast in polar region and slow in the disk, and the inclination is $i = 30^o$.

Two component are visible in the profile. A strong narrow emission which is the disk contribution, and a broad pedestal which is formed in

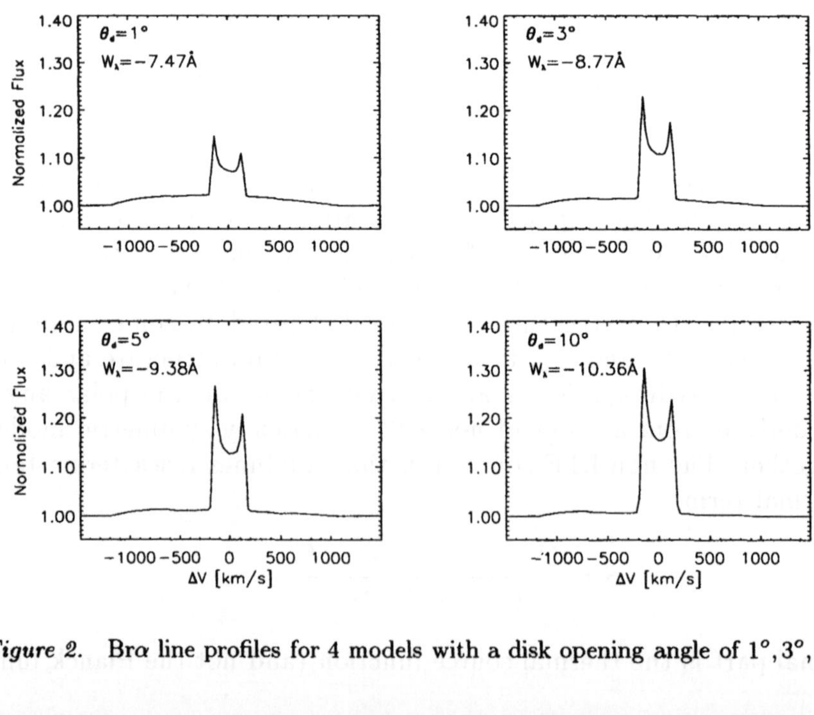

Figure 2. Brα line profiles for 4 models with a disk opening angle of $1^o, 3^o, 5^o$ and 10°.

the polar wind. The change in their relative contribution is clearly seen with increasing disk angle.

From this study and first ISO spectra of 4 B[e] stars, several preliminary results emerge. First, the apparent lack of a broad emission pedestal in ISO-SWS Brα observations indicates that the polar wind is relatively weak. Second, the hydrogen recombination lines are not resolved, confirming that velocities are small ($< 200\,km/s$) in the disk. This implies low terminal velocities and low inclination (or low rotation velocities). Finally, the Brα emission in MWC 922 and HD 127803 has a strength comparable to the model predictions. But the emission in CD -42° 11721 is stronger, pointing to a colder or a denser disk.

References

Bjorkman J.E., Cassinelli J.P., 1993, ApJ, 409, 429
Castor J.I., Dykema P.G., Klein R.I., 1992, ApJ, 387, 561
Ignace R., Cassinelli J.P., Bjorkman J.E., 1996, ApJ, 459, 671
Knigge C., Wood J.A., Drew J.E., 1995, MNRAS, 273, 225
Lamers H.J.G.L.M., Pauldrach A.W., 1991, A&A, 244, L5
Owocki S.P., Cranmer S.R., Gayley K.G., 1998, this volume
Poeckert R., Marlborough J.M., 1978, ApJ, 220, 940
Shlosman I., Vitello P., Mauche C.W., 1996, ApJ, 461, 377
Stee P., Araújo F.X., Vakili F., Mourard D., Arnold L., Bonneau D., Morand F., Tallon-Bosc I., 1995, A&A, 300, 219
Waters L.B.F.M., 1986, A&A, 162, 121
Zickgraf F.-J., 1998, this volume
Zickgraf F.-J., Wolf B., Stahl O., Leitherer C., Klare G., 1985, A&A, 143, 421

Discussion

Jaschek: 1 - You showed nicely that that you get double emission peaked profiles by inclination of the disk. Now what you observe is that in such structures V/R changes very often. Can you explain such changes?
2 - You have a thin disk and a large star. I guess one should see then the absorption spectrum of the star plus some emission lines. But what you see is a spectrum in which the absorption spectrum disappears at say 4200 Å and after that you see only emission lines. Can you explain this?
Lanz: 1- In an axisymmetric model, the V/R will primarily depend on inclination. To produce V/R variations in time, you would need some inhomogeneities in the disk or in the wind.
2 - The point where the absorption photospheric spectrum disappears depends on model parameters. If you see only emission lines in the optical it certainly means that the disk is larger i. e. the opening angle might be 10 or 15 degrees, rather than 3. Here I assumed a thin disk, as derived from the wind-compressed-disk model.

B[e] SUPERGIANTS: WHAT IS THEIR EVOLUTIONARY STATUS?

N. LANGER
Institut für Physik, Universität Potsdam
Postfach 601553, D–14415 Potsdam, Germany

AND

A. HEGER
Max-Planck-Institut für Astrophysik
D–85740 Garching, Germany

Abstract. In this paper, we investigate the evolutionary status of B[e] stars from the point of view of stellar evolution theory. We try to answer to the question of how massive hot supergiants — i.e. evolved stars — can be capable of producing a circumstellar disk. We find and discuss three possibilities: very massive evolved main sequence stars close to critical rotation due to their proximity to their Eddington-limit, blue supergiants which have just left the red supergiant branch, and single star merger remnants of a close binary system. While the latter process seems to be required to understand the properties of the spectroscopic binary R4 in the SMC, the other two scenarios may be capable of explaining the distribution of the B[e] stars in the HR diagram. The three scenarios make different predictions about the duration of the B[e] phase, the time integrated disk mass and the stellar properties during the B[e] phase, which may ultimately allow to distinguish them observationally.

1. Introduction

In recent years, the so called B[e] supergiants (cf. Lamers, this volume) have emerged as a distinct class of massive stars, whose defining properties — a strong mid-IR excess, strong Balmer emission, and narrow permitted and forbidden low-excitation emission lines — can be explained by a slowly outflowing equatorial disk superimposed to a normal fast wind (cf. Zickgraf

A. M. Hubert and C. Jaschek (eds.), B[e] Stars, 235–252.

et al. 1996a). Although the number of these objects is small — the best statistics exists for the Magellanic Clouds, where about 15 B[e] supergiants have been found (Gummersbach et al. 1995) — it is comparable to the number of Luminous Blue Variables (LBVs; cf. Bohannan 1997), which are know to represent a key evolutionary phase of very massive stars (Langer et al. 1994, García-Segura et al. 1997, Langer et al. 1998). It is thus of fundamental importance to understand whether the B[e] supergiants are just freaks, i.e. peculiar objects which come to exist due to special circumstances — in which case they might still show interesting physical phenomena — or whether all stars within a certain initial mass range evolve through a B[e] phase. In this case, our general understanding of the evolution of stars in that mass range may depend on our understanding of the B[e] stars.

Although we have a solid knowledge of the evolution of the deep interior of massive stars since a long time (Weaver et al. 1978, Kippenhahn & Weigert, 1990) it has become more and more clear during the last decades that our understanding of the evolution of their observable features is still rather incomplete. Among others, open problems concern the effective temperature evolution of moderately massive ($M_{ZAMS} \simeq 10...30\,M_{\odot}$) post main sequence stars (Langer & Maeder 1995) and the evolutionary connections between O stars, Luminous Blue Variables (LBVs) and Wolf-Rayet (WR) stars for higher masses (Schaller et al. 1992, Langer et al. 1994, Stothers & Chin 1996, Pasquali et al. 1997; see also Schulte-Ladbeck 1998). Two major physical difficulties in the theoretical models of the observable evolutionary stages of massive stars have been identified and made responsible for the persisting lack of reliable models: mass loss and internal mixing processes (Meynet et al. 1994, Langer 1994, Deng et al. 1996). E.g., it has been found that the mass loss of massive main sequence stars should be roughly twice as high as what appears to be observed in order to understand many features of massive post main sequence stars (Meynet et al. 1994, Langer et al. 1994). Additionally, there is growing evidence that stellar rotation may considerably affect the evolution of massive stars (Maeder 1987, Langer 1991a, Fliegner et al. 1996, Maeder & Meynet 1996, Meynet & Maeder 1997, Langer et al. 1997ab). *Rapid rotation* can reduce the effective gravity in the star, and it produces large scale flows (Eddington 1925). During the evolution, *differential rotation* occurs in all stars, with the possibility of the occurrence of various local hydrodynamic instabilities (cf. Endal & Sofia 1978, Zahn 1983) and corresponding mixing of chemical elements and angular momentum. Of relevance for massive stars are the shear instability (cf. Maeder 1997), the baroclinic instability (Zahn 1983, Spruit & Knobloch 1984), and the Solberg-Høiland and Goldreich-Schubert-Fricke instabilities (cf. Korycansky 1991).

Time dependent evolutionary models for massive stars including rotation have been constructed in the past in one dimension, using various degrees of approximation (e.g. Endal & Sofia 1978, Maeder 1987, Langer 1991a, Langer 1992, Talon et al. 1997, Langer 1998; cf. also Dupree 1995). Today, it is beyond reasonable doubts that the evolution of massive stars is influenced by rotation due to the physical mechanisms mentioned above (cf. Fliegner et al. 1996). While the principle effects of rotation in the interior of massive stars during their evolution all the way to iron core collapse are described elsewhere (Langer et al. 1997b, Heger et al. 1998), we discuss here whether massive single stars can approach the limit of critical rotation. We shall assume in Sections 2 and 3 that this would lead to the disk structure responsible for the B[e] phenomenon, either due to the Bjorkman-Cassinelli (1993) mechanism of wind compression of a rotating star or otherwise. A detailed wind disk model for B[e] stars can be found in Bjorkman (this volume), while the recent status of the Bjorkman-Cassinelli model is discussed by Owocki et al. (1996), Owocki (this volume) and Cassinelli (this volume). In Section 4 we will discuss a model where the disk is not produced by a critically rotating single star but rather a critically rotating close binary.

2. Very massive main sequence stars

Massive main sequence stars are rapid rotators, with equatorial rotation velocities in the range of 100...400 $km\,s^{-1}$ (Fukuda 1982, Penny 1996, Howarth et al. 1997). In our first scenario, we investigate whether massive main sequence stars, i.e. massive stars during core hydrogen burning, are capable of arriving a critical rotation (cf. Langer 1998). The results of this Section have been obtained with a hydrodynamic stellar evolution code (cf. Langer et al. 1988, Langer 1991b), employing the OPAL opacities of Iglesias et al. (1992). We have used an outer boundary condition which takes the optical depth of the stellar wind selfconsistently into account within a grey approximation (Langer et al. 1994, Heger & Langer 1996).

We have computed the proximity of the star to the Eddington-limit, i.e. the Eddington factor $\Gamma = L/L_{edd} = \kappa L/(4\pi cGM)$ in the following way. It has been shown in Langer (1997) that the occurrence of convection and of density inversions makes the concept of the Eddington limit as a stability limit invalid in the stellar interior, i.e. the Eddington factor Γ has to be evaluated only at the stellar surface. Since the term "surface" is not unambiguously defined in this context, we considered Γ in layers with an optical depth of $\tau < 100$, where in fact neither a significant convective energy flux nor density inversions have been found in the investigated models. Thus, to estimate the distance to the Eddington limit, we used the maximum value of Γ occurring in the subsonic layers with $\tau < 100$. Furthermore, we used the OPAL opacity coefficient to compute the Eddington factor. Mass

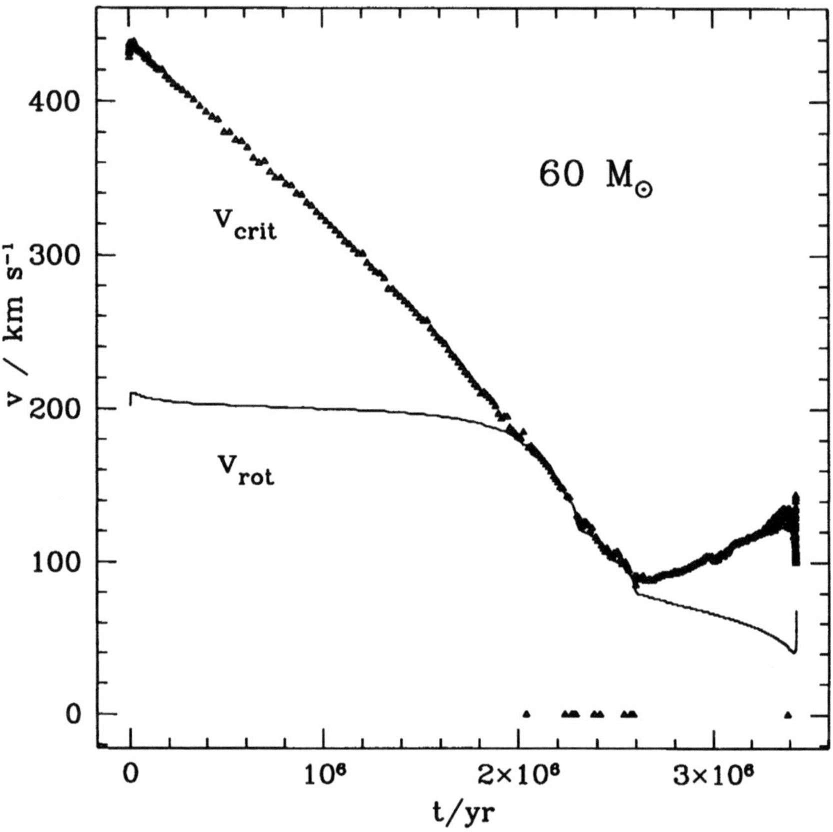

Figure 1. Equatorial rotational velocity v_{rot} and the critical rotational velocity v_{crit} as function of time during the main sequence evolution of a 60 $M_\odot$ sequence.

and luminosity are practically constant for $\tau < 100$ and equal to the total stellar mass and luminosity.

Angular momentum is carried only as a passive quantity in the stellar models, i.e. we ignore the centrifugal force in the stellar interior as well as the effect of mixing of chemical species due to rotationally induced instabilities (cf. Fliegner et al. 1996, Meynet & Maeder 1997). However, we do consider the centrifugal force at the stellar surface to evaluate the distance of the star from the Ω-limit (Langer 1997), i.e. from critical rotation, with $\Omega = v_{rot}/v_{crit}$, and $v_{crit}^2 = GM(1-\Gamma)/R$. Furthermore, we assume our models to be always rigidly rotating. Stellar models including differential rotation (e.g. Fliegner et al. 1996) show that this is a good approxima-

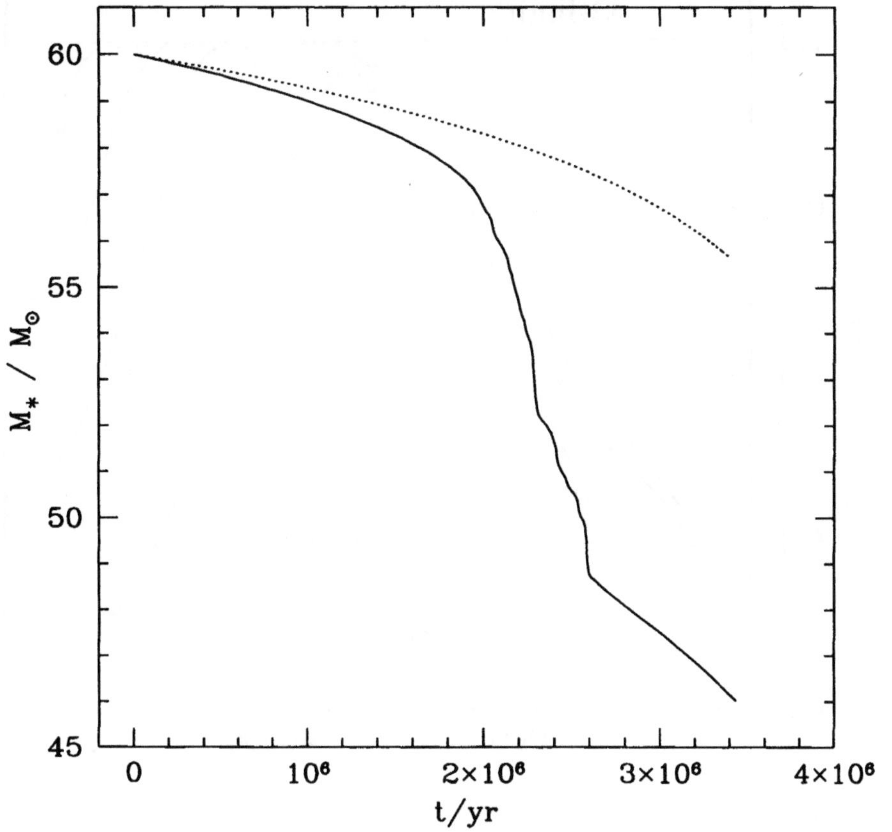

Figure 2. Stellar mass as a function of time for the 60 $M_\odot$ sequence shown in Fig. 1 (solid line), in comparison to the evolution of a 60 $M_\odot$ star without enhancement of the mass loss rate due to rotation (dotted line). The mass loss rates can be read of this figure as the slope of the curves.

tion on the main sequence, since the time scale for angular momentum transport is of the order of the thermal time scale and also shorter than the time scale of rotationally induced chemical mixing (Chaboyer & Zahn 1992, Zahn 1992, Talon & Zahn 1997). According to Zahn (1994), the expected amount of differential rotation in a massive main sequence star is roughly $\Delta\omega/\omega \simeq \omega^2 R^3/(GM)$, with ω being the mean angular velocity and $\Delta\omega$ its difference between stellar core and surface. In the models presented below, this estimate gives $\Delta\omega/\omega \lesssim 0.01$. The approximation of rigid rotation may become invalid when the mass loss time scale becomes shorter than that of angular momentum transport, but this is not the case in our models.

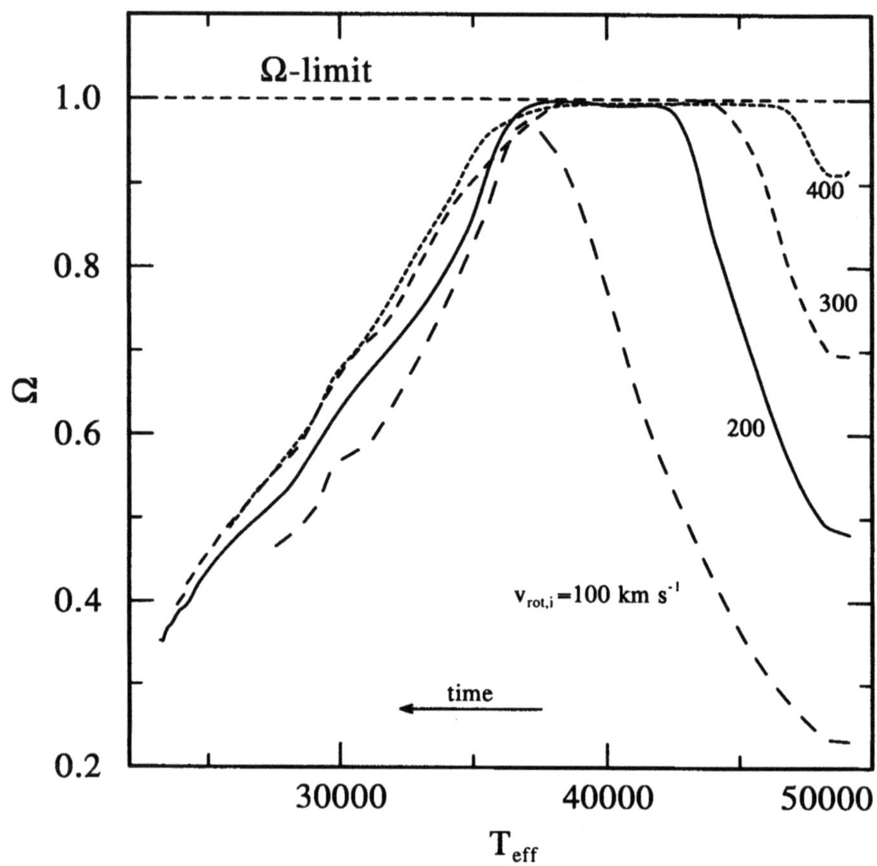

Figure 3. Evolution of the ratio of the rotation rate to the critical rotation rate, Ω, as function of the effective temperature during core hydrogen burning, for $60\,M_\odot$ sequences starting with different equatorial rotational velocities $v_{rot,i}$.

Angular momentum loss is only considered through the effect of mass loss, i.e. the lost mass carries away its specific angular momentum. To compute the mass loss rate for our stellar models, we have applied the empirical rate found by Lamers & Leitherer (1993) with the metallicity dependence obtained by Leitherer & Langer (1991). However, we have applied the correction factor derived by Bjorkman & Cassinelli (1993) as a fit to the results of Friend & Abbott (1986) to take the effect of rotation on the mass loss rate of hot star winds into account (cf. Langer 1998).

Fig. 1 shows the time evolution of the critical rotational velocity and the actual rotational velocity for a $60\,M_\odot$ sequence starting with $v_{rot} = 200\,km\,s^{-1}$. The critical rotational velocity has a pronounced minimum at roughly $t = 2.6\,10^6\,$yr, which corresponds to a stellar effective temperature of $T_{eff} \simeq 36\,500$ K, around which the iron opacity peak has its maximum effect. We see in Fig. 1 that, as v_{rot} approaches v_{crit} (i.e. $\Omega \rightarrow 1$) the rotation rate of the star declines such that the Ω-limit $\Omega = 1$ is never exceeded.

The reason is that, using the results of Friend & Abbott (1986), the mass loss rate increases as the star approaches $\Omega = 1$. In Section 3, we discuss that the angular momentum loss rate is directly coupled to the mass loss rate of the star (cf. Langer 1998, and Fig. 6 below). Consequently, the mass loss rate at the Ω-limit is determined by the angular momentum loss rate which is required to prevent the star from reaching $\Omega = 1$.

The result is that the star evolves along the Ω-limit until the critical rotation rate increases again, due to changes in the photospheric parameters (i.e., the opacity). For the considered example, the time spent at the Ω-limit is roughly $6\,10^5$ yr. Fig. 2 shows that during this time the mass loss rate is increased to roughly $10^{-5}\ M_\odot\ yr^{-1}$, which is a factor of ~ 10 above the normal radiation driven mass loss rate (cf. Fig. 2).

According to this picture, very massive main sequence stars may thus, due to their proximity to their Eddington-limit, evolve at critical rotation for several 10^5 yr. Fig. 3 shows that, for a given star, this time scale depends on its initial rotation rate. As during the time at the Ω-limit the star may have a slow dense equatorial outflow, a relationship to the B[e] stars may be suggested.

3. Supergiants on a blue loop

In this Section we describe a possibility for a star to arrive at critical rotation which does not require an extremely high stellar luminosity. It was found to occur in contracting stars, whose envelope structure changes from convective to radiative (see Heger & Langer 1998). This situation occurs for massive stars undergoing a so called blue loop, i.e. core helium burning red supergiants which evolve off the Hayashi-line towards the regime of blue supergiants in the HR diagram. Blue loops are typically found in the initial mass range 5...25 $M_\odot$ (cf. Langer 1991b, Schaller et al. 1992). However, the mechanism discussed in the following may also apply to other situations, e.g. to the post-AGB phase of low mass stars.

Fig. 4 shows evolutionary tracks for rotating stars between 10 and 20 $M_\odot$ (cf. Heger et al. 1997 for details), two of which evolve through a blue loop. The 12 $M_\odot$ sequence will be investigated in more detail below. Figure 5 shows the time evolution of the radii of various Lagrangian mass shells and of the stellar radius, before and during the blue loop. Also shown is the radial extent of the convective part of the envelope. We see from Fig. 5 that, within a fair approximation, the inner and the outer radius of the convective envelope remain constant until the blue loop occurs, while

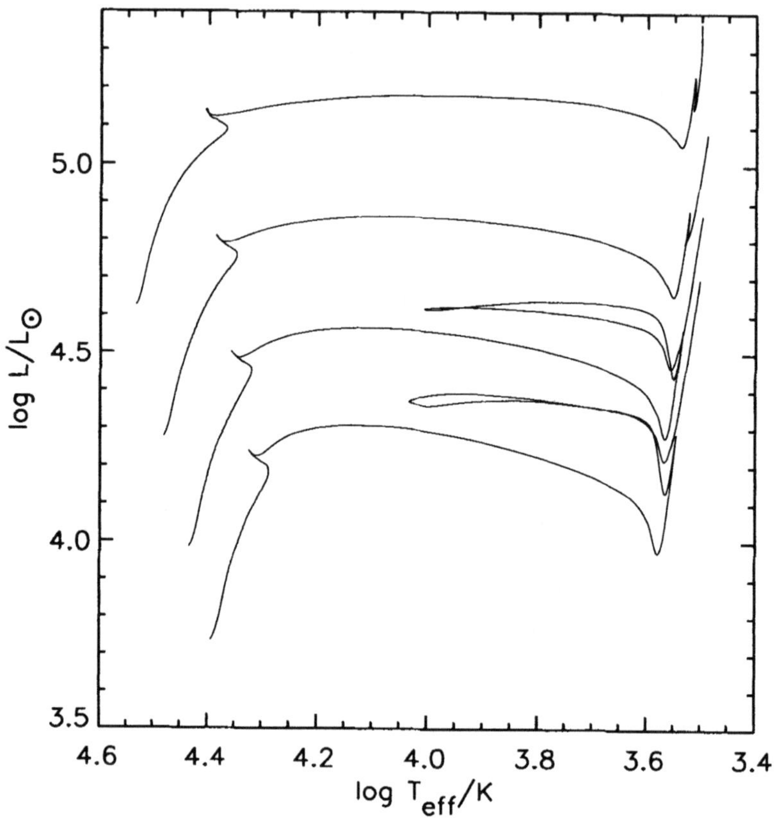

Figure 4. Evolutionary tracks for rotating 10, 12, 15 and 20 $M_{\odot}$ sequences from the ZAMS to central neon exhaustion (cf. Heger et al. 1997).

mass shells continuously flow out of the convection zone through its lower boundary.

In Fig. 6 we show that, due to the rapid angular momentum transport within convection zones, this situation leads to a dramatic spin-up of the envelope. In the right part of Fig. 6 (B1...B4), the situation is sketched for the assumption of rigid rotation in convective regions and broken up into three discrete steps. For a constant inner radius of the convection zone, the drop-out of mass through its lower boundary leads to an increase of the rotation frequency and of the mean specific angular momentum in the convection zone. Note that rigid rotation is assumed only for simplicity; it is not necessary for the spin-up mechanism to work.

The left side of Fig. 6 (A1...A4) sketches the situation of the loss of mass through the upper boundary of a rigidly rotating region. In this case,

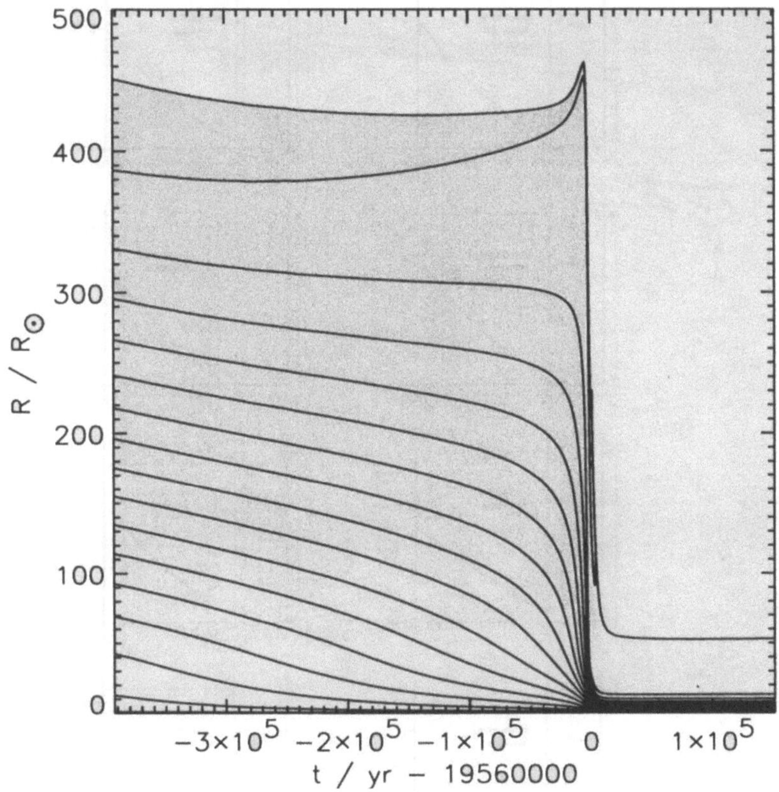

Figure 5. Evolution of the radii of different mass shells as a function of time for a period including the transition from the red to the blue supergiant stage of our 12 $M_\odot$ model. $t = 0$ is defined as in Fig. 7 and corresponds roughly to the time of the red-blue transition. Except for the uppermost solid line, which corresponds to the surface of the star, the lines trace Lagrangian mass coordinates. The mass difference between the lines is 0.5 $M_\odot$. Shading indicates convective regions.

which applies to the mass loss of very massive main sequence stars (cf. Sect. 2), matter with the highest specific angular momentum is continously lost, which leads to a spin-down of the rigidly rotating star.

Fig. 7 shows the time dependence of the rotational velocity of the 12 $M_\odot$ sequence before, during, and after the red-blue transition. The transition itself takes only about 10 000 yr. During this time the angular momentum transport to the surface layers of the star continues. The envelope layers remain rigidly rotating, continuing to shovel up angular momentum to the surface. The contraction of the star by a factor of $f \approx 10$ would increase the rotational velocity by the same factor if j were conserved locally (cf.

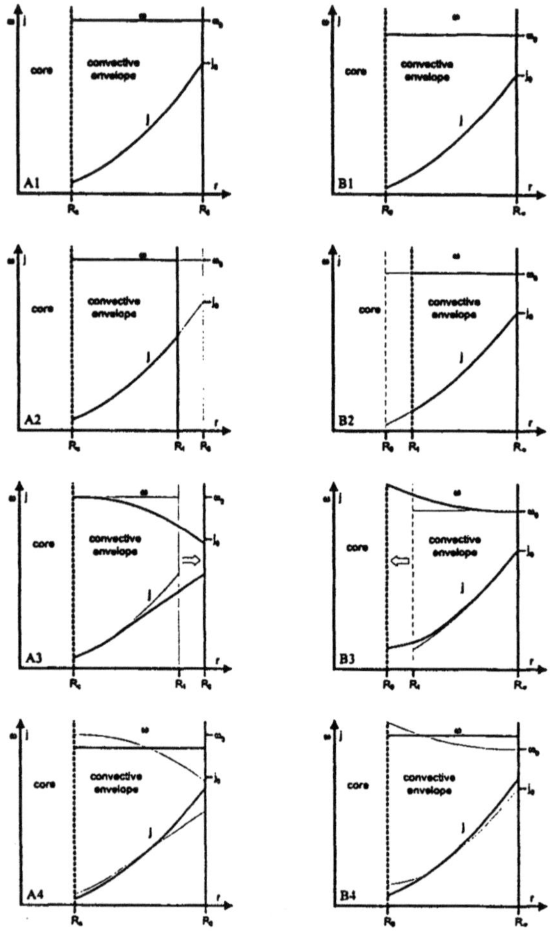

Figure 6. Mass loss from a rigidly rotating stellar envelope from the surface (case A: left panels) and through its lower boundary (case B: right panels). The continuous process is split up into three steps. First (panels $1 \rightarrow 2$), mass gets lost from the envelope, secondly, the envelope restores its original (inner or outer) radius R_0 (panels $2 \rightarrow 3$) by expansion, and third (panels $3 \rightarrow 4$) the specific angular momentum j is redistributed such that rigid rotation (i.e. $\omega(r)$ =const.) is restored. This leads to spin-down (spin-up) and decrease (increase) of the mean specific angular momentum for the case of mass loss through the upper (lower) boundary of the rigidly rotating stellar envelope. Thin lines show the state of the preceding step.

Fig. 7, "decoupled").

Since $1 - \Gamma \not\ll 1$, the Keplerian angular velocity $\omega_{\rm Kep} = \sqrt{GM/R^3} \propto f^{3/2}$ can be used as a rough approximation for the critical rotation frequency $\omega_{\rm c}$. Thus, Ω scales as $f^{1/2}$, and the mere contraction, without any angular momentum transport, would bring the star by a factor ~ 3 closer to its critical rotation rate. The numerical value of $\Gamma \approx 2\,10^{-3}$ on the RSG found

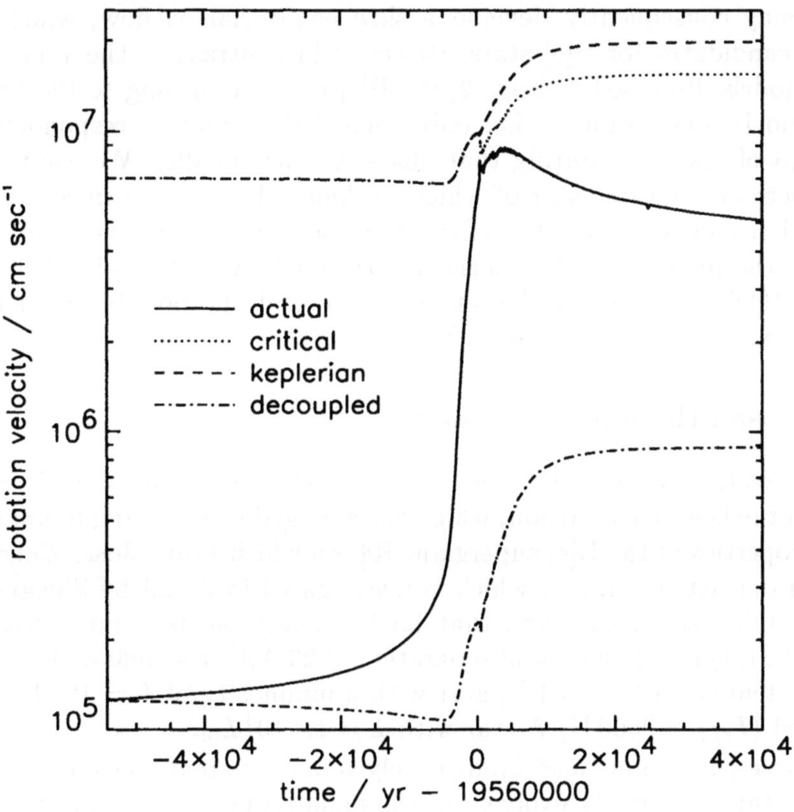

Figure 7. Equatorial rotation velocity as a function of time (solid line) compared to the Keplerian (dashed line) and the critical (dotted line) rotation rate; the latter two are different by the factor $1-\Gamma$. During the red supergiant phase it is $\Gamma \ll 1$ and the two lines coincide, while during the blue supergiant phase Γ rises to 0.4. The dash-dotted line shows the evolution of the surface rotation rate if there were no angular momentum transport in the convective envelope.

in our calculation and $\Gamma \approx 0.4$ on the BSG would lead to an increase of Ω by ~ 3.9 from the RSG to the BSG, again assuming that j were conserved locally. Actually, the star gets from $\Omega \approx 0.01$ on the red supergiant branch at the beginning of central helium burning to critical rotation ($\Omega = 1$) during the RSG/BSG transition, even though a considerable loss of angular momentum due to stellar wind mass loss is included in our model (cf. Fig. 7).

In fact, the model sequence would have exceeded the Ω-limit if we would not have applied a mass loss increase for $\Omega \to 1$ as in the massive main sequence models described in Section 2. I.e., like in that case, we find that stars undergoing a blue loop are likely to arrive at critical rotation, and they may consequently develop a slow equatorial outflow, which makes them candidates for B[e] stars. However, in contrast to the very massive star models discussed in Sect. 2, the B[e]-phase according to the blue loop scenario is much shorter, i.e. only some 10^4 yr, and correspondingly the amount of mass lost during that phase is much smaller. We want to mention here at least one star of which we know that it performed a red-blue evolution and for which the effect described here almost certainly played a role: the progenitor of Supernova 1987A (cf. Arnett et al. 1989, Langer et al. 1989; cf. also Woosley et al. 1997), and the possible twin of it as described by Brandner et al. (1997).

4. R4 and the binary scenario

In this Section, we want to investigate a binary scenario as possible explanation of the B[e]-phenomenon, which has emerged while trying to understand the properties of the B[e] supergiant R4, on which Franz-Josef Zickgraf has drawn our attention, and which is investigated in detail by Zickgraf et al. (1996b). These authors find that R4 is a spectroscopic binary with a period of 21.3 yr and an orbital separation of 23 A.U. ($\sim 5000\,R_\odot$). Presently, the system consists of a B[e] star with a luminosity of $L = 10^5\,L_\odot$ and an evolved ($T_{eff} \simeq 9500$ K) A star with $L = 1.4\,10^4\,L_\odot$.

These parameters lead immediately to an apparent contradiction: since the A star has already evolved off the main sequence, the almost 10 times more luminous companion should have become a supernova long time ago. Also, the orbital separation is too large in order to solve this contradiction by assuming that the A star was once more massive and shed mass onto the B component.

However, another binary scenario may be able to not only explain the properties of both stars but also the existence of a disk around the B star. In this scenario, the system was initially a triple system consisting of a close pair of stars with an initial mass of $\sim 10\,M_\odot$ for each of them, and another $\sim 10\,M_\odot$ star surrounding the pair on a wide orbit. The latter star is identical to the present A star, and only serves as a clock without ever interacting with the close pair.

The evolution of the close pair is sketched in Fig. 8. In order to end the evolution with the merging of both stars, the more massive component — the primary — must lose mass to the secondary at a high rate by Roche-lobe overflow through the first Lagrange-point L_1. If the accretion time scale $\tau_{\dot{M}} = M/\dot{M}$ of the secondary is shorter than its thermal time scale $\tau_{KH} = GM^2/(RL)$ it will swell and start filling its own Roche-lobe. As the expansion will not stop here, matter will soon leave the system through the

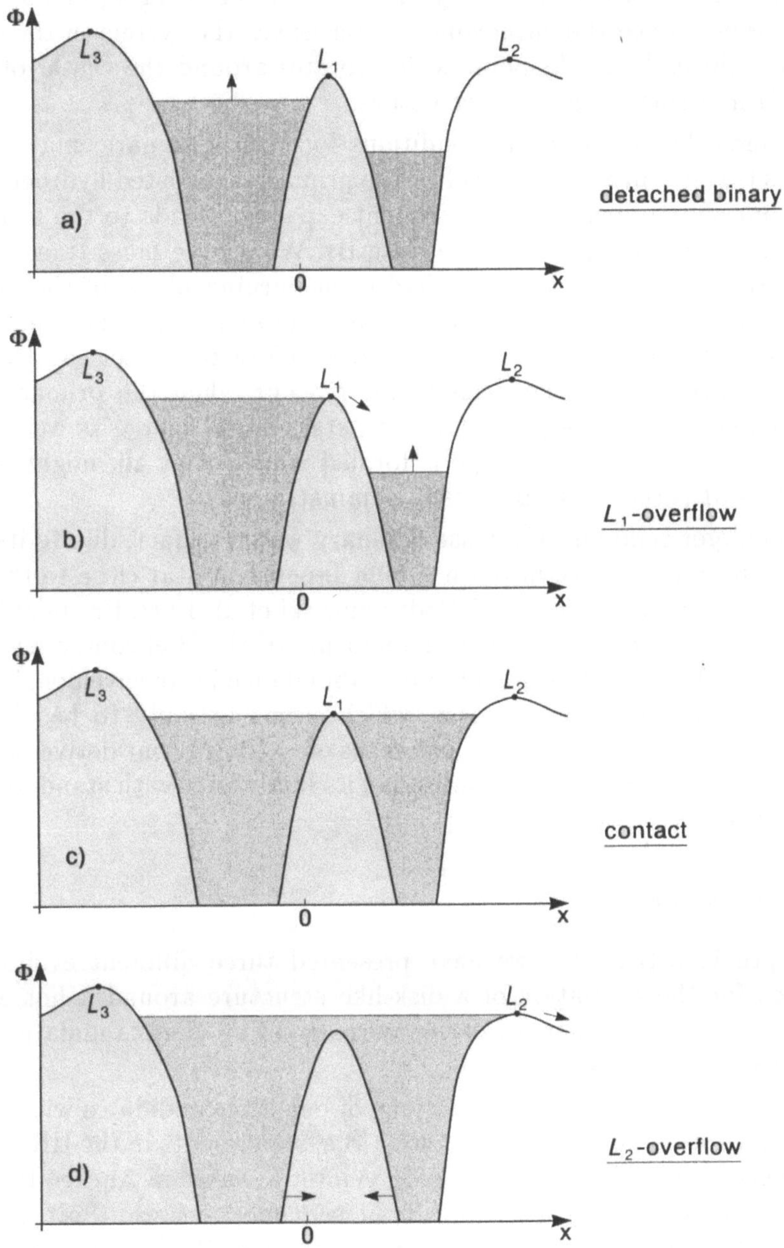

Figure 8. Schematic evolution of the spatial extents of two stars of a close binary system which evolves through a contact phase and develops mass outflow through the second Lagrange-point ("L_2-overflow"), which can be responsible for the formation of a circumsystem disk or ring (see text for details).

second Lagrange-point L_2, and the corresponding angular momentum loss will bring both stars continuously closer, with the merging of both stars as final result. Since the outflowing matter leaves the system in the orbital plane, i.e. through the L_2-point which rotates around the center of mass, it may form a ring or disk-like structure.

The most favorable initial conditions for such a scenario may be that the mass transfer process starts after the primary exhausted hydrogen in its center, but shortly before its subsequent expansion leads to the formation of a convective envelope (i.e., late Case B). Would the mass transfer start much earlier, i.e. during the core hydrogen burning phase of the primary (Case A), the merger star, which is now much more massive than the distant companion, would evolve into a supernova before the latter can leave the main sequence. Would the mass transfer occur when the primary has a red supergiant structure (Case C), both stars might merge as well, but so quickly that a common envelope is formed which, if at all, might tend to leave the star rather in a spherically symmetric way.

The merger remnant of a Case B binary would in fact, due to its small helium core mass, evolve into a hot blue supergiant star close to the main sequence in the HR diagram (cf. Podsiadlowski et al. 1992, Braun & Langer 1995), which fits to the effective temperature of the B[e] component of R4 of 27 000 K. Due to a strong helium overabundance in its envelope, it would appear overluminous for its mass, which seems actually to be observed: Zickgraf et al. (1996b) find a present mass of $\sim 13\,M_\odot$, but derive an initial mass of $\sim 20\,M_\odot$ from the comparison of its luminosity with standard single star evolutionary tracks.

5. Conclusions

In the previous Sections, we have presented three different evolutionary scenarios for the formation of a disk-like structure around a hot evolved massive star. The properties of the corresponding B[e] candidate models are compared in Table 1.

First, we can compare the positions of our B[e] candidates with the observed distribution of Magellanic Cloud B[e] supergiants in the HR diagram (Gummersbach et al. 1995; cf. also de Winter & van den Ancker 1997, and Zickgraf, this volume). Clearly, our main sequence scenario (Sect. 2) would not work for the stars with luminosities below $\sim 10^5\,L_\odot$. However, those stars are found to be comparatively cool ($\log T_{eff} \simeq 4.1$) so that they fit quite well to the blue loop scenario outlined in Sect. 3. The very luminous group of B[e] stars investigated by Gummersbach et al., which can not correspond to stars on a blue loop, are mostly rather hot ($\log T_{eff} \simeq 4.4$) and might well be main sequence stars. I.e., both scenarios together cover

TABLE 1. Comparison of observable properties predicted by the three B[e] evolutionary scenarios discussed in this paper.

	very massive main sequence star at the Ω-limit	supergiant on blueward excursion from Hayashi line	single star remnant of binary merger
T_{eff}	ok for most luminous B[e]s	ok for less luminous B[e]s	scatter in T_{eff} plausible
luminosity	ok for most luminous B[e]s	ok for less luminous B[e]s	scatter in L plausible
time scale	some 10^5 yr	some 10^4 yr	(?)
time integr. disk mass	$\sim 5\,M_\odot$	$\sim 0.1\,M_\odot$	$\sim 5\,M_\odot$(?)

roughly the range of observed B[e] supergiants in the HR diagram.

For the binary scenario (Sect. 4), there are no quantitative models, but it appears likely that, assuming a scatter in the initial orbital parameters and in the initial mass ratio, the observed scatter in the HR diagram could also be reproduced.

As indicated in Table 1, the time scale of the B[e] phenomenon and the time integrated equatorial mass loss are rather different for the three models. While the main sequence model predicts a long B[e] life time with several solar masses expelled through the disk wind, the blue loop scenario predicts a shorter life time and less equatorial mass loss. The expected number of B[e] stars from both scenarios may still be comparable due to the steep decline of the initial mass function for larger masses.

B[e] time scale and disk mass loss are least clear for the binary scenario. During the L_2-overflow phase (cf. Fig. 8) certainly of the order of several solar masses of matter are lost. However, it is unclear which fraction of that is pushed to infinity and which fraction forms a disk. Another possibility which might be considered is that the star which is formed by the merger process is an extremely rapid rotator, and consequently has an equatorially focussed wind. In this case, the time scale of the B[e] phenomenon would be of the order of the spin-down time scale of the star.

Finally, we may also consider to combine the three scenarios discussed above. For example, if the merger of two stars is a red supergiant, e.g. in a Case C system, it may have considerably more angular momentum than red supergiant formed from single stars. If such a star evolves into a blue supergiant, our blue loop scenario must operate again, with an even stronger effect than in the single star case worked out in Section 3. It may be interesting to note in this context that Podsiadlowski (1997) suggested such a merger scenario for the presupernova evolution of SN 1987A.

Acknowledgements

The authors are very grateful to Peter Conti, Guillermo García-Segura, André Maeder, Stan Owocki, Philipp Podsiadlowski, Regina Schulte-Ladbeck, and Franz-Josef Zickgraf for enlightening discussions, and to Lars Koesterke for major help in the production of figures of this paper. This work has been supported by the Deutsche Forschungsgemeinschaft through grant No. La 587/15-1

References

Arnett W.D., Bahcall J.N., Kirshner R.P., Woosley S.E., 1989, ARAA 27, 629
Bjorkman J.E., Cassinelli J.P. 1993, ApJ 409, 429
Bohannan B., 1997, in *Luminous Blue Variables: Massive Stars in Transition*, A. Nota, H.J.G.L.M. Lamers, eds, ASP Conf. Ser., Vol. 120, p. 3
Brandner W., Grebel E.K., Chu Y.-H., Weis K., 1997, ApJ 475, L45
Braun H., Langer N., 1995, A&A 297, 483
Chaboyer B., Zahn J.-P., 1992, A&A 253, 173
Deng L., Bressan A., Chiosi C., 1996, A& A 313, 159
Dupree R.G., 1995, ApJ 439, 357
Eddington A.S., 1925, Observatory 48, 73
Endal A.S., Sofia S., 1978, ApJ 220, 279
Fliegner J., Langer N. & Venn K., 1996, A& A 308, L13
Friend D.B., Abbott D.C., 1986, ApJ 311, 701
Fukuda I., 1982, PASP 94, 271
García-Segura G., Langer N., Mac Low M.-M., 1997, in *Luminous Blue Variables: Massive Stars in Transition*, A. Nota, H.J.G.L.M. Lamers, eds, ASP Conf. Ser., Vol. 120, p. 332
Gummersbach C.A., Zickgraf F.-J., Wolf B., 1995, A&A 302, 409
Heger A., Langer N., 1996, A& A 315, 421
Heger A., Langer N., 1998, A&A, in preparation
Heger A., Jeannin L., Langer N., Baraffe I., 1997, A&A 327, 224
Heger A., Woosley S.E., Langer N., 1998, ApJ, in preparation
Howarth I.D., Siebert K.W., Hussain G.A.J., Prinja R.K., 1997, M.N.R.A.S. 284, 265
Iglesias C.A., Rogers F.J., Wilson B.G., 1992, ApJ 397, 717
Kippenhahn R., Weigert, A., 1990, Stellar Structure and Evolution, Springer, Berlin
Korycansky D.G., 1991, ApJ 381, 515
Lamers H.J.G.L.M., Leitherer C., 1993, ApJ 412, 771
Langer N., 1991a, A&A 243, 155
Langer N., 1991b, A&A 252, 669
Langer N., 1992, A&A 265, L17
Langer N., 1994, in Proc. 34th Herstmonceux Conference on *Circumstellar Media in the Late Stages of Stellar Evolution*, R. Clegg et al., ed., Cambridge University Press, p. 1
Langer N., 1997, in *Luminous Blue Variables: Massive Stars in Transition*, A. Nota, H.J.G.L.M. Lamers, eds, ASP Conf. Ser., Vol. 120, p. 332
Langer N., 1998, A&A 329, 551

Langer N., Maeder A. 1995, A&A 295, 685
Langer N., Kiriakidis, M., El Eid, M.F., Fricke, K.J., Weiss, A., 1988, A&A 192, 177
Langer N., El Eid, M.F., Baraffe I., 1989, A&A 224, L17
Langer N., Hamann W.-R., Lennon M., Najarro F., Pauldrach A.W.A., Puls J., 1994, A& A 290, 819
Langer N., Heger A., Fliegner J., 1997a, in proc. IAU-Symp. 189 on *Fundamental Stellar Properties: The Interaction between Observation and Theory*, T. R. Bedding, A. J. Booth and J. Davis, eds., Kluwer, Dordrecht, p. 343
Langer N., Fliegner J., Heger A., Woosley S.E., 1997b, Nucl. Phys. A, 621, 83c
Langer N., García-Segura G., Mac Low M.-M., 1998, ApJ, submitted
Leitherer C., Langer N., 1991, in: IAU-Symp. 148 on *The Magellanic Clouds*, R.F. Haynes, D.K. Milne, ed., Kluwer, p. 480
Maeder A., 1987, A&A 178, 159
Maeder A., 1997, A&A 321, 134
Maeder A., Meynet G., 1996, A& A 313, 140
Meynet G., Maeder A., 1997, A& A 321, 465
Meynet G., Maeder A., Schaller G., Schaerer D., Charbonnel C., 1994, A&AS 103, 97
Owocki S.P., Cranmer S.R., Gayley K.G., 1996, ApJ 472, L11
Pasquali A., Langer N., Schmutz W., Leitherer C., Nota A., Hubeny I., Moffat A.F.J., Drissen L., Robert C.: 1997, ApJ 478, 340
Penny L.R., 1996, ApJ 463, 737
Podsiadlowski P., 1997, in: *SN 1987A: Ten years after*, Eds. M.M. Phillips, N.B. Suntzeff, PASP, in press
Podsiadlowski P., Joss P.C., Hsu J.J.L., 1992, ApJ, 391, 246
Schaller G., Schaerer D., Meynet G., Maeder A. 1992, A&AS, 96, 269
Schulte-Ladbeck R., 1998, this volume
Schulte-Ladbeck R., Clayton G.C., 1993, AJ 106, 790
Spruit H.C., Knobloch E., 1984, A&A 132, 89
Stothers R.B., Chin C.-W., 1996, ApJ 468, 842
Talon S., Zahn J.-P., 1997, A& A 317, 749
Talon S., Zahn J.-P., Maeder A., Meynet G., 1997, A&A 322, 209
Weaver T.A., Zimmerman G.B., Woosley S.E., 1978, ApJ 225, 1021
de Winter D., van den Ancker M.E., 1997, A&AS 121, 275
Woosley, S.E., Heger, A., Weaver, T.A., Langer, N., 1997, in: *SN 1987A: Ten years after*, Eds. M.M. Phillips, N.B. Suntzeff, PASP, in press
Zahn J.-P., 1983, in proc. 13th Saas-Fee Course, A. Maeder et al., eds., Geneva Observatory
Zahn J.P., 1992, A&A, 265, 115
Zahn J.P., 1994, in: *Evolution of Massive Stars: A Confrontation between Theory and Observations*, D. Vanbeveren et al., eds., Kluwer, Dordrecht, p. 285
Zickgraf F.J., Humphreys R.M., Lamers H.J.G.L.M., Smolinski J., Wolf B., Stahl O., 1996a, A&A 315, 510
Zickgraf F.J., Kovács J., Wolf B., Stahl O., Kaufer A., Appenzeller I., 1996b, A&A 309, 505

Discussion

Owocki: Your scaling for mass loss approaching the Ω-limit is based on the Friend & Abbott result that assumes a uniformly bright stellar surface. With gravity darkening, such mass loss formula still predicts a larger mass loss, but from the poles. Now, even if you assume that the radiation fed in at the base of the outflow is not gravity darkened, then it is still unlikely that the radiation flux will remain constant through the disk, as required to maintain an equatorial outflow, since radiations scattered from the disk will tend to escape at higher latitide. Finally, the opacity increase from the iron bump should not remain as the density and temperature decrease outward. This suggests that any flow will stagnate. From a point of view of dynamics, it seems very difficult to radiatively drive such an equatorial outflow without spining material up to Kelperian orbit first.

Langer: At present, we do not know how and if gravity darkening works in a massive star with convective outer layers; Kippenhahn (1977, A&A 58, 267), who gegeralized von Zeipel's results from the 20ies, finds that even "gravity brightening" (i.e. that the radiative flux increases with decreasing effective gravity) can occur. And your suggestion that **any** radiation driven disk-like outflow has to stagnate unless Kelperian rotation velocities are obtained is disproven by your own work (ApJ 424, 887; 1994). I think that we might both agree that the underlying physics involved in such flows needs more investigations.

Bjorkman: Regarding Stan Owocki's comment, whether or not there should be any gravity darkening in stars with convective envelopes: do you know if there is any difference in the optical depth from the top of the convective zone (to $r = \infty$) between the equator and the pole? If not then it seems that there should be little difference in the ability of photons to diffuse out the equator relative to higher latitudes.

Langer: I'm afraid this depends on the temperatures and densities which are established in the equatorial region, which are unknown a priori.

Zorec: Have you taken into account any angular momentum transport inside the star? Calculations by Kippenhahn, Sofia, etc. showed that a star starting as a rigid rotator on the main sequence rotates rigidly in the main sequence but then becomes differential rotator with a higher specific angular momentum in the core. Conti et al. showed that O stars seem to be differential rotators. Zorec et al. have also shown that the same phenomenon may exist among B stars.

Langer: Angular momentum transport in the stellar interior is considered in our models. We do not find strong differential rotation on the main sequence, but in the supergiant stage we do.

Cassinelli: I want to follow up on my comment this morning and ask about SN 1987A. Could there be a red supergiant merger that would later lead to a B[e] star?

Langer: This is quite possible. I have concentrated in my talk on Case B merger, simply since we understand the physics better in such events, but Case C merger (i.e. involving a red supergiant) can not be discarded (see above, and Podsiadlowski 1997).

EVOLUTIONARY CONNECTIONS AMONG MASSIVE STARS – AN OBSERVER'S PERSPECTIVE

REGINA E. SCHULTE-LADBECK
University of Pittsburgh, Dept. of Physics & Astronomy, Pittsburgh, PA 15260

Abstract.
B[e] supergiants are considered to be the evolutionary descendants of rapidly rotating massive stars. While B[e] supergiants populate a region of the HRD which overlaps with that of other massive stars, such as normal AB supergiants and Luminous Blue Variables (LBVs), it has yet to be established how they fit in with evolutionary scenarios for massive stars. I propose several observational tests.

1. The upper HRD – how do B[e] supergiants fit in?

While it is well accepted that massive OB stars end their lives in supernova explosions, the exact evolutionary sequence(s) are as yet unknown. Maeder (1997 and references therein), for example, published an elaborate scheme for the (theoretical) evolution of massive stars, for different initial masses and chemistries. But, presently none of the available stellar-evolution models includes the B[e] supergiants as a phase in the evolution of massive stars (Schulte-Ladbeck 1997).

In analyzing the B[e] component of the binary R4, Zickgraf et al. (1996) tentatively found that C is underabundant, and N overabundant, with respect to solar, and that the stellar mass is now only about 50% of the ZAMS mass. These findings suggest a post-main-sequence stage for B[e] supergiants (a clear distinction from similar objects, such as the Herbig AeB[e] stars, also discussed at this conference).

A major difficulty in ascertaining the evolution of massive stars is that they do not exhibit (in most cases) atmospheric absorption lines. Therefore, elaborate model atmospheres codes are needed to derive stellar parameters. Figure 1 illustrates that the most luminous B[e] stars overlap in tempera-

A. M. Hubert and C. Jaschek (eds.), B[e] Stars, 253–258.

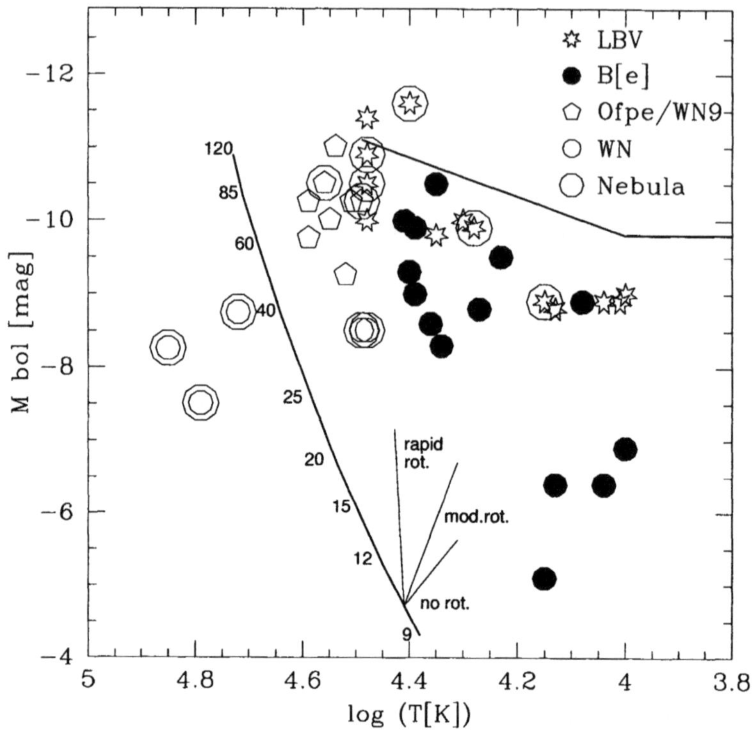

Figure 1. The location of massive stars of several galaxies on the HRD. Data for LBVs (in quiescence) and the HD limit are from Humphreys & Davidson (1994), for Ofpe/WN9 stars from Pasquali et al. (1997), and from Gummersbach (1994) for B[e] stars. Only those WN star which show ejecta nebulae are plotted; the stellar parameters are from Hamann et al. (1995). The presence of ring nebulae is also indicated, and is based on Smith (1995) and Nota & Clampin (1997). Absence of ring nebulae in the B[e] class is due to lack of observations. The location of the ZAMS for solar metallicities was taken from Schaller et al. (1992). The direction of an evolutionary track of a 10 solar mass star without and with rotation from Fliegner et al. (1996) is also shown.

ture and luminosity range with LBVs. It is well established that WR stellar atmospheres are H deficient; the spectral types WN and WC are considered pre-supernova stages of massive stars. Crowther (1997) summarizes results for H/He surface abundances; the similar values for LBVs, normal B supergiants, and WN11 stars suggest an evolutionary connection. With the exception of the analysis of the R4 binary, virtually nothing is known about the surface abundances of B[e] supergiants. An opportunity for observers presents itself with those B[e] stars which we view along the rotation axis. Pole-on B[e] stars show, as best we know, a "normal" hot-star atmosphere, hence their spectra could be modeled with existing hot-star atmosphere codes – a model for the disk need not be included.

Curiously enough, the well-known LBV Eta Car is on the original B[e] star list of Allen & Swings (1976). The fact that it's spectrum doesn't exhibit [OI] lines may have caused it to be removed from lists of B[e] stars after Zickgraf et al. (1985) introduced spectroscopic criteria for the B[e] supergiant class. Hillier (1992) and Hamann et al. (1994) comment on similarities of Eta Car with the B[e] stars, such as the presence of a disk and of hot circumstellar dust.

A few observations are known which suggest connections between the stellar types in the upper HRD. In order to become an LBV, a luminous, blue star must *vary*, and either show a >3 mag "giant eruption" (which has been observed directly for only two stars in this class, P Cyg and Eta Car), or it must at least show "cyclic LBV variability" with an amplitude of 1 to 2 mag (characterized by a change in temperature at constant luminosity), which, to the best of our knowledge, has a recurrence time of decades. It is immediately obvious that observational bias is a major problem in establishing the LBV character of massive stars.

The LMC star R 127 showed LBV-type variability (Stahl 1983) and was originally thought to be of Ofpe/WN9 type in quiescence, but has recently been re-classified as WN 11 (Crowther 1997). The galactic star AG Car displays LBV-type variations, and was classed by Stahl (1986) as an Ofpe/WN9 star in quiescence, but this was revised to a WN 11 type by Smith et al. (1994). Morris et al. (1996) point out that the K-band spectrum of AG Car in its latest LBV-type event resembled very much that of B[e] stars.

There is a lingering misconception in the community, due to some early work of Zickgraf et al., that B[e] stars are not variable. Many authors have pointed out spectroscopic and polarimetric variability in the B[e] supergiants S22 (e.g., Shore 1992, Schulte-Ladbeck & Clayton 1993) and S18 (Schulte-Ladbeck et al. 1994b, Morris et al. 1996). The most direct link between B[e] and LBV stars comes from the observation that the B[e] binary R 4 in the SMC recently underwent a variation which showed characteristics of LBV-type variability (Zickgraf et al. 1996).

In the absence of knowledge on their variability histories, massive stars have been advanced to LBV-type candidacy if they show circumstellar nebulae. Kinematically associated nebulae containing 1-2 solar masses of processed material are thought to be the fossil records of major eruptions. Unfortunately, while observations of nebulae around LBVs and Ofpe/WN9 stars are readily available (e.g., Nota & Clampin 1997), *there has not yet been a systematic survey for extended nebulosity around B[e] supergiants.* In the case of the SMC B[e] star N82, Heydari-Malayeri (1990) suspected associated emission nebulosity based on direct images, but couldn't confirm it with long-slit spectroscopy. The fact that one B[e] supergiant doesn't exhibit a ring nebula need not discourage further study – not all LBVs show them either.

The two-component wind model proposed by Zickgraf et al. (1985, 1986) indicates rotation must be an important parameter for B[e] supergiants. The circumstellar disks even provide a location to shield CO molecules and dust from the UV radiation of the star. While only one LBV shows CO emission (HR Car) and one exhibits an IR excess due to hot dust (Eta Car), several LBVs show atmospheric density-asymmetries (e.g., Schulte-Ladbeck et al. 1994 a,b) and are surrounded by axisymmetric ejecta nebulae (Nota et al. 1995). This indicates that rotation (or binarity) has played a role in the evolution of LBVs as well. The polarization characteristics of LBVs, B[e] stars and WR stars have led Schulte-Ladbeck (1995) to propose that of the OB stars evolving off the main sequence, the most rapidly rotating stars will be able to retain angular momentum into stages as advanced as WN. Therefore, in addition to initial mass and chemical composition, *rotation is another fundamental parameter in the evolution of massive stars.* Langer (1997 & this conference) recently discussed evolution with rotation, which could involve a B[e] phase prior to the LBV phase. Evolution with rotation (see Fig. 1) makes the association of a particular location on the HRD with a particular evolutionary track impossible – a rapidly rotating massive star might masquerade as a star of even higher mass on a non-rotating track.

2. Conclusions and suggestions for future work

Our inability to establish the evolutionary stage of B[e] supergiants is due to a lack of observational data. With the data in hand, one is led to conclude that B[e] stars are in a post-main-sequence phase (R4 shows enriched abundances and has a reduced mass). Their evolutionary state could be either pre-LBV (N82 does not exhibit an associated nebula indicative of a giant eruption) or young LBV (if we count Eta Car among the B[e] supergiants) or pre-RSG (in the case that some B[e] stars avoid the LBV phase altogether). In order to establish the evolutionary connection of B[e] stars with other massive stars, the following observations are key:

- a systematic search for circumstellar nebulae
- surface abundance analyses
- continued photometric and spectroscopic monitoring.

Acknowledgements

This work was supported in part by NASA grant GO-06693.02-95A through STScI.

References

Allen, D.A., Swings, J.P., 1976, A&A, 47, 293
Crowther, P., 1997, ASP Conf. Ser., 120, 51
Fliegner, J., Langer, N., Venn, K.A., 1996, A&A, 308, L13
Gummersbach, C.A., 1994, Diplomarbeit, University of Heidelberg
Hamann, F., DePoy, D.L., Johansson, S., Elias, J., 1994, ApJ, 422, 626
Hamann, W.-R., Koesterke, L., Wessolowski, 1995, A&A, 299, 151
Heydari-Malayeri, M., 1990, A&A, 1990, 234
Hillier, D.J., 1992, in "The Atmospheres of Early-Type Stars", Lecture Notes in Physics, 401, 105
Humphreys, R.M., Davidson, K., 1994, PASP, 106, 102
Langer, N., 1997, ASP Conf. Ser., 120, 83
Maeder, A., 1997, ASP Conf. Ser., 120, 374
Morris, P.W., Eenens, R.P.J., Blum, R.D., 1996, ApJ, 470, 597
Najarro, F., Hillier, D.J., Stahl, O., 1997, preprint
Nota, A., Clampin, M., 1997, ASP Conf. Ser., 120, 303
Nota, A., Livio, M., Clampin, M., Schulte-Ladbeck, R.E., 1995, ApJ, 448, 788
Pasquali, A., Langer, N., Schmuts, W., Leitherer, C., Nota, A., Hubeny, I., Moffat, A.F.J., 1997, ApJ, 478, 340
Schaller, G., Schaerer, D., Meynet, G., Maeder, A., 1992, A&AS, 96, 269
Schmutz, W., Leitherer, C., Hubeny, I., Vogel, N., Hamann, W.-R., Wessolowski, U., 1991, ApJ, 372, 664
Schulte-Ladbeck, R.E., 1995, IAU Symp., 163, 176
Schulte-Ladbeck, R.E., 1997, RvMA, 10, 135
Schulte-Ladbeck, R.E., Clayton, G.C., 1993, AJ, 106, 790
Schulte-Ladbeck, R.E., Clayton, G.C., Hillier, D.J., Harries, T.J., Howarth, I.D., 1994a, ApJ, 429, 864
Schulte-Ladbeck, R.E., Clayton, G.C., Leitherer, C., Drissen, L., Robert, C., Nota, A., Parker, J. Wm., 1994b, SSRv, 66, 193
Shore, S.N., 1992, ASP Conf. Ser., 22, 342
Smith, L.J., 1995, IAU Symp. 163, 24
Zickgraf, F.-J., Wolf, B., Stahl, O., Leitherer, C., Klare, G., 1985, A&A, 143, 212
Zickgraf, F.-J., Wolf, B., Leitherer, C., Appenzeller, I., Stahl, O., 1986, A&A, 163, 119
Zickgraf, F.-J., Kovacs, J., Wolf, B., Stahl, O., Kaufer, A., Appenzeller, I., 1996, A&A, 309, 505

Discussion

Zickgraf: For R 4 estimates for CO abundances exist (overabundant, C underabundant). This object also showed, at least during one epoch, one LBV brightness outburst.

Schulte-Ladbeck: I did mention your result on R 4. However since it was derived by comparing R 4's spectrum to Kurucz model atmospheres, who knows what it really means. Therefore I still think that we know very little about the surface abundances of B[e] stars and that abundances are the most important piece of information needed to clarify the evolutionary state of B[e] stars. I was not aware of your observations of an LBV-like emission in R 4.

Friedjung: I presume that if you wish to find abundances you should look out for UV spectra. There you should see the central star without too much disturbance by disk emission.

Schulte-Ladbeck: I suggest to begin with the pole-on B[e] star to derive all stellar parameters but rotational velocity. Viewed from above the poles, through their fast, low-density winds, it should be easy to apply the same type of analysis and modeling used for other hot stars and thus to derive a consistent set of stellar parameters which can be used to compare the evolutionary state of B[e] stars and other hot stars with stellar winds.

Viotti: Chemical abundances are difficult to derive from the stellar spectrum, but can be determined from the circumstellar nebulae, as it has been made for η Car and AG Car. For instance from the analysis of the nebulae around the WO star WR 102 (Sand 4) we (Polcaro et al. 1994) found a large gradient in nitrogen and the abundances, which might trace the past history of the star after having left the main sequence.

Schulte-Ladbeck: Indeed. However we do not know of any ring nebulae around B[e] stars at this time. If we did, the kind of analysis which you propose would be very valuable to establish the evolutionary history of B[e] stars. Therefore I suggested as a promising line of future work on B[e] stars a systematic search for circumstellar nebulae.

Cassinelli: Your comment that we see no ring nebulae around B[e] stars seems to rule out the possibility that B[e] progenitors are red supergiants. I say this because red supergiants tend to have very massive slow winds. Therefore these winds should get swept up by supergiant winds to form spherical shells.

Schulte-Ladbeck: Thank you. I will add that to my paper. I guess in preparing the talk I was prejudiced by the Fliegner et al track of evolution going UP in the HR diagram and did not think sufficiently about observations that might suggest or rule out a red supergiant phase.

THE NATURE OF THE EXTREME EMISSION LINE STAR MWC 297

RENÉ D. OUDMAIJER, JANET E. DREW, GRAEME BUSFIELD

Imperial College of Science, Technology and Medicine,
Blackett Laboratory, Prince Consort Road,
London, SW7 2BZ, U.K.

1. Introduction

MWC 297 is one of the objects in Herbig's (1960) original list of young stars, which were later called the Herbig Ae/Be stars. Since the object displays a prodigious infrared excess, forbidden lines in the spectrum, and an early spectral type (based on photometry), MWC 297 satisfies all the criteria of being a B[e] star, and it appears regularly in catalogues containing B[e] stars. Because of its large reddening, MWC 297 has so far evaded a spectroscopic spectral classification, and consequently, the distance and nature of the object remained to some extent unclear. In this contribution, we use high signal-to-noise (SNR > 150), medium resolution (0.8) blue spectra, with a total exposure time in excess of 8 hours, that we have obtained at the William Herschel Telescope in June 1994, and will show that MWC 297 is a rapidly rotating B1.5Ve star located in the Aquila Rift.

2. Its spectral type and rotational speed

Fig. 1 shows the co-added spectrum around 4600 Å together with data obtained for three early-B stars of known spectral type. The value of es-

A. M. Hubert and C. Jaschek (eds.), B[e] Stars, 259–262.

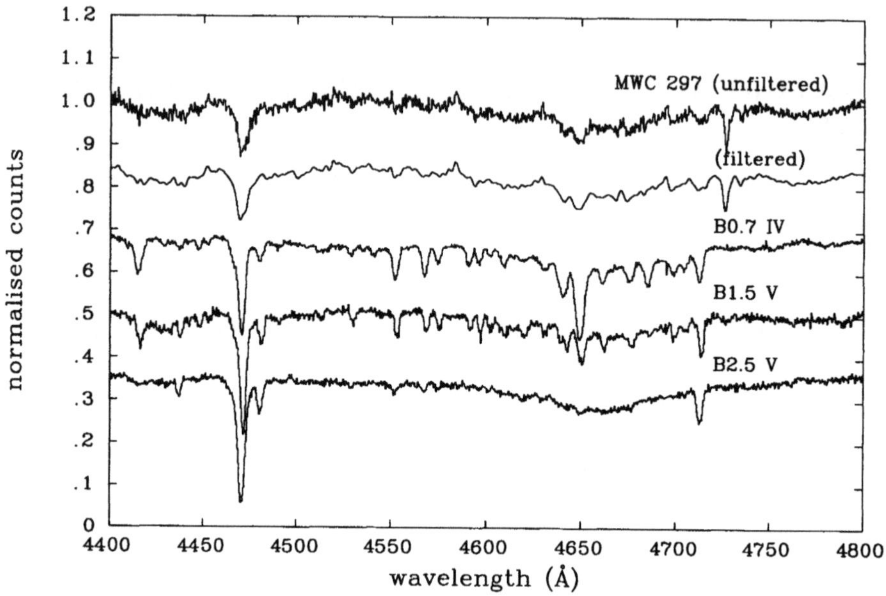

Figure 1. The normalised spectrum of MWC 297 is compared with similarly normalised early B-type spectra. In the top row, the data on MWC 297 are shown unsmoothed. In the second row, the same data are shown after applying a low-pass Fourier filter to eliminate high frequency noise. The B-star spectra are are unsmoothed. The broad depressions at either end of the wavelength range in MWC 297 are plausibly DIBs as is the prominent, quite narrow feature at 4726 Å.

pecially this wavelength region for spectral-typing purposes is that there are noticeably spectral-type dependent heavy element lines to be found in between the He I 4471Å and 4713Å features. Of particular note are the features at ~4640Å and ~4650Å which are blends of O II, N III and C III lines. It is immediately apparent that if any of the weaker heavy element lines present in their spectra are also present in MWC 297, they must be substantially broader in order to appear at such low contrast. MWC 297 resembles the B1.5 star most, but the large rotational broadening of its lines makes a good spectral classification difficult. We therefore chose to individually rotationally-broaden the standard star spectra in steps of $v \sin i$ and then to compare the r.m.s. deviation between the spectrum of MWC 297 and the standard stars (see Drew et al. 1997 for more details). The best match is found with the B1.5 object with its lines broadened to an equivalent $v \sin i$ of 350 ± 50 km.s^{-1}. To illustrate this method, Fig. 2 shows all standard stars convolved with $v \sin i = 300$ km.s^{-1} (in the case of the B1.5 star, this corresponds to a total line broadening of 350 km.s^{-1}), with MWC 297 overlaid on them. Clearly, the best match is for the B1.5 star.

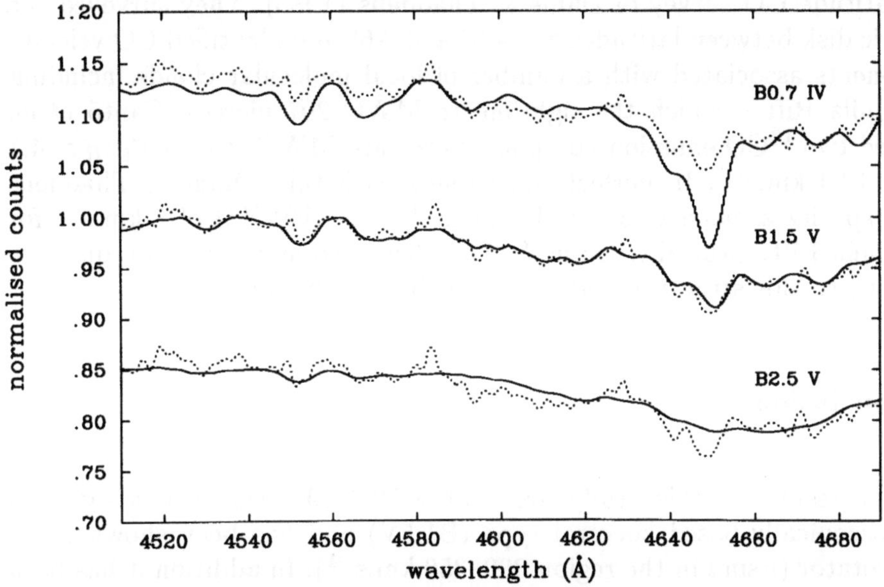

Figure 2. The Fourier filtered spectrum of MWC 297 (dotted line) is shown superposed on each of the comparison B stars (solid lines) after the latter have been convolved with a $v \sin i = 300$ km.s^{-1} theoretical line profile. This shows clearly that the best match is between MWC 297 and the B1.5V star, HD 154445. Peaks at 4549 Å, 4583 Å and 4629 Å in MWC 297's spectrum are likely to be extremely weak Fe II emission.

3. Its location in the Galaxy

In the recent literature, the distance to MWC 297 has been estimated to be about 500 pc. For example Bergner *et al.* (1988) proposed 530 ± 70 pc (and a B0 spectral type). The not dissimilar figure of 450 pc was suggested by Cantó *et al.* (1984) who obtained it from CO velocities measured along its sightline. The latter noted however that kinematic estimates cannot be relied upon when the implied distance is much under 1 kpc, as in this case. As discussed above, the spectral type of MWC 297 is B1.5 with an uncertainty of half a spectral class. In addition, the star appears to be a rapid rotator ($v \sin i$ > 300 km s^{-1}). This makes it most plausible that MWC 297 is on or very close to the main sequence. If indeed the object is a B1.5V star, its absolute visual magnitude can be estimated to be −2.8, using Schmidt-Kaler (1982). On adopting a visual extinction, A_V= 8 (see Drew *et al.* 1997), and a mean apparent *V* magnitude of 12.2 (Bergner *et al.* 1988), we arrive at the much more modest distance of 250 ± 50 pc. Strong support for this smaller distance is provided by the results of the wide-latitude CO survey of Dame & Thaddeus (1985). They surveyed the Galactic disk between latitudes $b = -5°$ and $+6°$ and identified CO velocity components associated with a number of local molecular clouds including the Aquila Rift – which the sightline to MWC 297 pierces. Cantó *et al.* detected two ^{13}CO emission components toward MWC 297, at $V_{LSR} \simeq 4.1$ and $\simeq 12.4$ km.s^{-1}. In perfect consistency with this, Dame & Thaddeus (1985) specify a mean velocity of 8 km.s^{-1} and a FWHM of 7 km.s^{-1} for the Aquila Rift. Crucially, Dame & Thaddeus also assign a mean distance of 200 pc to this structure and an upper limit of 300 pc.

4. Conclusions

The key results of this study are that MWC 297 has been assigned a spectroscopically-based spectral type (B1.5V) and has been shown to be rapid rotator ($v \sin i$ in the region 300–350 km.s^{-1}). In addition it has been argued that commonly quoted distances to the object are too large by around a factor of two – a distance revised downwards to 250 ± 50 pc allows an association between MWC 297 and the molecular gas of the Aquila Rift. Therefore, since MWC 297 satisfies all the criteria normally required for classification as a Herbig Be star (location, emission line spectrum, IR excess) it is the most rapidly rotating object of its type yet recognised. Its spectral type is the earliest spectral type determined spectroscopically for a Herbig Ae/Be star. As MWC 297 also shows characteristics of B[e] stars, it seems reasonable to follow the nomenclature proposed by Lamers at these proceedings, and call MWC 297 a Herbig B[e] star.

We end with the note that by virtue of its relatively modest reddening MWC 297 can be considered a unique opportunity to study the link between the lower mass, less obscured Herbig Ae/Be stars and the higher mass, optically invisible massive Young Stellar Objects, the BN-type stars.

References

Bergner Yu. K., *et al.*, 1988, Astrofizika, 28, 529
Cantó J., Rodríguez L. F., Calvet N., Levreault R.M., 1984, ApJ 282, 631
Dame T. M., Thaddeus P., 1985, ApJ 297, 751
Drew J.E., Busfield G., Hoare M.G., Murdoch K.A., Nixon C.A., Oudmaijer R.D. 1997, MNRAS 286, 538
Herbig G. H., 1960, ApJS, 4, 337
Schmidt-Kaler Th., 1982, in Landolt-Börnstein, Vol. 2b. Springer-Verlag, Berlin

THE SPECTROSCOPIC VARIETY OF DIFFERENT CATEGORIES OF B–EMISSION LINE STARS

R. VIOTTI[1], G. MURATORIO[2], V.F. POLCARO[1]
1. Istituto di Astrofisica Spaziale, CNR, Frascati, Italy
2. Observatoire de Marseille, Marseille, France

AND

C. ROSSI[3], M. VILLADA[4]
3. Istituto Astronomico, Università La Sapienza, Roma
4. Observatorio Astronomico UNC, Cordoba, Argentina

Abstract. Emission lines are frequently observed, but poorly studied, in early–type stars. We discuss which information can be derived from a more systematic study of the emisison lines (including line profiles and variability) in a number of selected representatives of the different categories.

1. Some spectroscopic characteristics

The appearence of emission lines in a stellar spectrum generally provides indirect evidence for the presence of extended stellar atmospheres, stellar winds, and circumstellar disks/nebulae, while the connection with the physical structure of the underlying star still is matter of discussion. In spite of many decades of studies, the potential of information which can be in principle derived from the emission lines, has only marginally been used. One reason is the large variety of emission line spectra seen, which imply different approaches. A way to overcome the problem would be to concentrate our studies on the analysis of the spectra of a few "good representatives" of the different categories, the blue spectrum of some of which are reproduced in Fig.1. It should be noticed in the figure the richness and strength of the Fe II emission lines in LBV prototype η Car, in the LMC supergiant S22 and in the Herbig Ae/Be V380 Ori. Note also the variety in line width and profile in the different objects. The Stark–broadened Balmer lines are visible in the lower luminosity stars. He I which is present in absorption in

A. M. Hubert and C. Jaschek (eds.), B[e] Stars, 263–267.

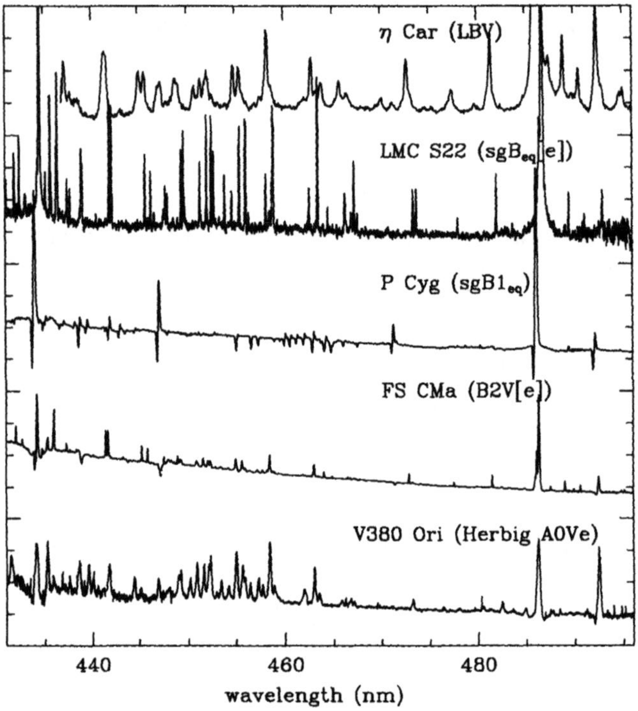

Figure 1. The blue spectra of some representative emission line stars obtained at OHP–1.52m (P Cyg, FS CMa, V380 Ori), CASLEO–2.15m (η Car), and ESO–1.52m (S22).

FS CMa, is present (and variable) in emission in the much cooler object V380 Ori.

The richness of the emission line spectrum in some stars is illustrated by the many different species identified in the optical spectrum of these two stars:

FS CMa (B2V[e]): H I, Na I, Si II, Mg II, Ti II, Cr II, Mn II, Fe II, [N II], [O I], [S II], [Cu II], [Fe II], [Co II], [Ni II].

V380 Ori (A0Ve): H, He I, O I, Na I, Ca II, Mg II, Ti II, Cr II, Sc II, V II, Fe II, [Fe II].

This is suggestive of a statistical treatment of the different ions – such as the SAC method described by M. Friedjung in this Volume – in order to determine the relative ion abundances. For instance, the richness of the Ti II lines in V380 Ori seems not to be due to a titanium overabundance in this star, but rather to a combined effect of level excitation and line opacity. In fact, Rossi et al. (1997) found that once these effects are taken into account, the Ti^+/Fe^+ abundance ratio is probably lower than cosmic.

The study of the *line profiles* can tell us about the geometry and dynamics of the emitting regions (rotating equatorial disks, expanding atmospheres, stellar winds). Viotti et al. (this Volume) have shown that in the near-IR spectrum of the classical Be star γ Cas the Pδ line has a double-horn structure, with the blue horn weaker than the red one, suggesting that the matter moving towards us is somewhat absorbed, or that there is some geometrical asymmetry. The Hα profile observed at the same time was completely different (Viotti et al. 1997), because of the different line opacity.

2. Evolutionary stage

An important question concerns the evolutionary stage of emission line stars, which can be investigated in different manners. (a) From the *stellar environment*, if the star belongs to a stellar association of known distance and age. We have for instance studied the case of the young association Berkeley 87 which includes several early type stars with Hα in emission (Polcaro & Viotti 1997). (b) From the *chemical composition* and dynamics of the circumstellar matter. This is illustrated by the case of the ring nebula of the LBV star AG Car where N is more abundant with respect to the outside matter which was ejected in a previous state (perhaps red supergiant) of the star. In the nebula surrounding the evolved WR/WO star Sand 4, we have found a trend of the He and N abundance which might be tracing the previous WN and LBV stages of this massive star (Polcaro et al. 1995). (c) The determination of the *chemical abundance* of the emission line spectrum of the star itself is another key for understanding the nature of the emission line stars. However, our systematic study of these objects has unveiled that in most cases emission lines are strongly optically thick which prevents a direct determination of the relative abundances, e.g. using an emission line curve of growth analysis (see M. Friedjung, this Volume).

3. Cautions and needs

In treating the spectrum of such peculiar stars one should take care of a number of possible errors and misunderstandings:

- *Shell/wind absorption.* Absorption lines may be formed in the stellar envelopes (disks, shells, winds). At low resolution they cannot be easily distinguished from the photospheric lines, and frequently the "envelope spectrum" has been used – erroneously – for spectral classification. Typical examples are in the UV the high ionization resonance lines (N V, C IV, Si IV) formed in the stellar winds, and the wide 160–180 nm blend of singly ionized metal lines seen in some shell spectra as in the case of the Herbig Ae star V380 Ori (Rossi et al. 1997).

- *Spectral classification.* Spectral and luminosity classification helps the determination of the position of the star in the H–R diagram. But no classification is possible if there are no truly photospheric lines visible, which is the case of η Car, S22 and P Cyg in Fig.1. In these cases a more appropriate terminology would be to indicate the "equivalent" spectral type, when it is derived from the energy distribution (e.g. P Cyg: $B1_{eq}$), not from the ionization level of the emission line spectrum.
- *Time variability.* B–emission stars are known to be spectroscopic variables. For instance AG Car's effective temperature has varied between 10000 K and 30000 K in a few years (from $A1_{eq}$ to $O9_{eq}$). In some other cases the forbidden lines disappear. It is therefore a nonsense to put a star in a given category on the basis of a limited number of spectroscopic observations.
- *Line opacity.* Since most emission lines are strongly optically thick, their relative fluxes cannot be used for classification purposes. For the same reason, the Fe II/[Fe II] ratio is not a good (direct) diagnostic of the envelope density.

In conclusion, there is a number of requirements for a real progress in our understanding the nature of B–emission stars. (a) First, we need *high quality spectroscopic material* in order for an accurate measure of line intensity (or equivalent width) especially of the weakest emissions ($W_{eq} \leq 0.01$ nm). This will also overcome the problem of ambiguous line identification. (b) It is also urgently needed a careful monitoring of the *spectral variability* (with different time scales). This should necessarily be made for a limited number of selected objects which are considered good representatives of the different stellar categories. (c) Finally, it is important to *check for binarity* in order to select effects which are due to the interaction of the companion star rather than to the structure of the star itself which we are studying.

References

Polcaro V.F., Viotti R. (1997), 7th Annual ADASS Conference, ASP Conf Ser. in press
Polcaro V.F., Rossi C., Norci L., Viotti R. (1995), *A&A*, **303**, 211
Rossi, C., Errico, L., Friedjung, M., et al. (1997), *A&A*, submitted
Viotti R., Rossi C., Muratorio G. (1997), *A&AS*, in press

Discussion

Hubert: I would like to mention the relative difficulty to determine the spectral temperature of classical Be stars, compared to B[e] stars and from "photospheric" lines. A good example is provided by 31 Peg (Zorec et al,

1996, A&A 308, 852), with the study of the wings of the Hγ photospheric lines.
Jaschek: Please be cautious -HD 45677 may not be the typical B[e] star-it shows many forbidden lines which are not present in other stars of this group.
Viotti: I said that FS CMa is a good representative - also for being rather bright and for the many studies done so far on this star-not a typical member of an uncertain stellar category.
Vakili: There are recent interferometric observations of γ Cas in H_α evidencing axially asymmetries in the envelope of this (your?) prototype of Be stars (Sánchez Petrov, A&A submitted) that the authors interpret as the presence of a very close (inside the H_α envelope) low mass companion. I suggest to consider multiplicity as part of diagnostic studies of Be stars and to take ψ Per as the Be prototype.
Viotti: Historically, γ Cas is the prototype, but I agree that we might select another candidate(s) for a detailed study. From my point of view I prefer the star I am studying is not a close binary, since I am interested in the physics of outer envelopes of stars without "external" influences.
Friedjung: I insist that we need high resolution spectra, which simultaneously cover a large wavelength range. The objects are variable, so if different spectral lines are observed on different nights it is not clear whether they can be compared with each other.
Viotti: I agree. I have in fact compared the profile of P_δ and H_α of γ Cas and have taken care to say that they have been observed at the same time.

EELOS: CAN WE DEFINE A GALACTIC B[e] GROUP?

D. DE WINTER
Dpto. Física Teórica, C–XI, Universidad Autónoma de Madrid, Cantoblanco, 28049 Madrid, Spain

AND

M.R. PÉREZ
Space Applications Corp., Suite 250, 9315 Largo Drive West, Largo, MD 20774, USA

Abstract. The class of B[e] stars is a spectroscopically defined group from which members can be identified as LBVs, supergiants, symbiotics, post-AGB, proto-planetary and young stars. Objects that cannot be classified as such are frequently designated to be a special stellar class with an unidentified (evolutionary) status. To investigate if such a new stellar B[e] class holds an homogeneous stellar population we have searched for galactic B[e] stars among several samples of EELOs.

Based on the spectral definition alone, we find 31 B[e] stars and 18 candidates that are unclassified. We argue that most of these objects could be member of the classes mentioned when properly characterized and their astrophysical properties are better determined. We find only a few objects that are well studied and could not be classified. On statistical and evolutionary grounds we argue that these objects will end up in one of the known classes leaving no further arguments to define new classes of B[e] stars. Based on these results, we show and propose a new classification scheme for B[e] stars.

1. Introduction

A sample of extreme emission line objects (EELOs), of unknown evolutionary status, was first introduced in the catalogue of Herbig Ae/Be (HAeBe) candidates (Thé et al. 1994). The name EELO refers to the large number of emission lines in the optical spectrum similar of Bep or B[e] stars.

A. M. Hubert and C. Jaschek (eds.), B[e] Stars, 269–276.

Although, most of the identified EELOs are of spectral type B, the classification B[e] is only appropriate as a spectroscopic definition. B[e] stars form, therefore, a subgroup of EELOs. Reasons to use the term EELOs are that it refers to the existence of forbidden or permitted emission lines, whatever the evolutionary status or spectral type of an object might be. Therefore, this term does not define a small group of stars with peculiar spectra which can be easily miss-classified, when treated as an homogeneous group of stars as it is frequently and erroneously done for B[e] stars (see the many discussions in this proceedings). In other words, there are several diverse and exclusive stellar groups that contain B[e] stars, such as LBVs, B[e] supergiants, symbiotics, (P)PNs and HAeBes of which some of the members are spectroscopically B[e], but of course, not related at all.

It is this confusion that makes it very tempting to define (new) homogeneous groups of B[e] stars based on some (spectroscopic) characteristics. Such definitions are described by Lamers et al. (this proceedings) and will not be repeated here. In this contribution, we discuss the selection of B[e] among EELOs. The EELO samples used are themselves, selections of different databases. After the selection proposed, we present a complete list of yet unclassified B[e] stars with some of their observed properties (for which we have used yet unpublished low resolution spectroscopy and photometry). Finally, we present which homogeneous groups of galactic B[e] stars could be differentiated and whether new homogeneous groups can be defined in order to classify all B[e] stars identified so far.

2. Selection of B[e] stars and candidates

2.1. THE B[e] SELECTION CRITERIA

The selection criteria for B[e] stars is simple; the star must be of spectral type B, or its optical spectrum should indicate the presence of a hot stellar object, with the presence of emission lines such as [Fe II] and [O I]. We must note that some B[e] stars, e.g. LBVs, might undergo phases with very weak or absent [Fe II] lines. For completeness and in order to include some B[e] candidates, we also select objects with very strong Fe II and tracers of [Fe II] emission lines.

2.2. SAMPLES OF EELOS

The EELO samples used here are taken from the catalogue of emission lines in star forming (SF) regions of Cohen & Kuhi (1979) and from the catalogue of HAeBe candidates of Thé et al. (1994). We warn that our sample is strongly biased toward young stars, but this is indeed not a fatal selection effect if these young stars were selected strictly based on observable properties and not exclusively due to their location in SF regions. Both, Herbig Be and B[e] stars, can show similar characteristics such as forbidden and permitted emission lines, a near- and far-IR excess and both can be found in SF regions as well as field objects. Such a sample is selected from the catalogue of Thé et al. (1994).

2.2.1. *Sample A: Stars in SF regions*
The Cohen & Kuhi catalogue shows the spectra of 406 probable young Hα emission lines stars and it is in principle an unbiased sample with no preference toward early spectral types. Investigation of the published spectra, indicates the presence of emission lines of Fe II for 33% of the objects, 56% shows He I, and 33% have [O I] emission lines. This indicates that over 100 EELOs are among these young emission line stars. If we take a detailed look at the Fe II, [Fe II] and [O I] emission characteristics, we find the following statistics against spectral types:

Group ------>	I	II	III	
Emission lines:	O/B	?	K/M	Totals
moderate/weak Fe II	4	2	17	23
and strong [O I]	2	1	3	6
and/or [Fe II]	0	0	0	0
strong Fe II	3	10	4	17
and [O I]	3	9	2	14
and/or [Fe II]	1+1w	3w	2w	7

(w = weak emission)

It is of particular interest to note the lack of A/F/G-type stars with these strong emission characteristics. For the selection of B[e] stars, we are only interested in groups I & II. The [e] stars, with unclassified spectral type, are V510 Ori, HS Ori and RW Aur, the latter two being probably T Tauri stars (TTs). Together with the B[e] stars R Mon and V380 Mon, they form a class of young B[e] candidates.

2.2.2. *Sample B: Candidate Herbig Ae/Be stars*
The catalogue of Thé et al. (1994) lists 322 candidate HAeBe stars taken from a wide collection of other samples that fulfill their selection criteria such as the original HAeBe lists, samples of Be stars with near-IR excess, as in Allen & Swings (1976), and the sample of Dong & Hu (1991). The latter source is important as it is a selection of the large Wackerling (1970) cata-

TABLE 1. Non-supergiant and unclassified B[e] stars with optical Fe II & [Fe II] spectral lines

Object	Neb./MC/SF	Isolated	IRAS	NIRE	[O I]	Sp.T.	Remark
New:							
HK Ori	SF	(Y)	Y	Y	Y	G:[e]p	
V380 Ori	SF	(Y)	Y	Y	N	B9[e]+sh	
R Mon	SF	(Y)	Y	Y	Y	B0[e]	
He 3–847		N	Y	Y	Y	B6/A0:[e]	
He 3–1191	Neb.	Y	Y	Y	Y	B0:[e]	Post-AGB?
AS 119		Y	Y	Y	Y	B8/9[e]	
Th35–27	Neb.	Y	Y	Y	Y	B[e]	
Confirmed:							
CD–42°11721	Neb.	Y	Y	Y	Y	B0[e]p	[Fe II] var?
HD 45677		Y	Y	Y	Y	B3V[e]p+sh	
HD 50138		Y	Y	Y	Y	B6V[e]+sh	[Fe II] var?
HD 51585		N	Y	Y	Y	B[e]q	Symb?/Bin?
HD 87643	Neb.	Y	Y	Y	Y	B3/4[e]	
HD 326823	Neb./MC	Y	Y	Y	?	B1.5[e]	
SS 73–170		Y	Y	Y	Y	B[e]	
He 3–1138		Y	Y	Y	?	B8Ia[e]	LBV?
MWC 939		Y	Y	Y	Y	B[e]	[Fe II] var?
Allen & Swings:							
MWC 17		Y	Y	Y	Y	B[e]	Symb?/Bin?
MWC 84		Y	Y	Y	?	B[e]	
MWC 819	Neb.	Y	Y	Y	Y	B[e]	
He 3–52		Y	Y	Y	Y	B[e]	
He 2–79		Y	Y	Y	Y	B[e]	
He 2–91	Neb.	Y	Y	Y	Y	B[e]	Plan. neb?
AS 269	Neb.	Y	Y	Y	Y	B[e]	Symb?/Bin?
MWC 922		Y	Y	Y	Y	B[e]	
PN M 1–92	Neb.	Y	Y	Y	Y	B0.5IV[e]	
He 2–446	SF	(N)	Y	Y	Y	B:[e]	
MWC 623		Y	Y	Y	N	B[e]	Symb?/Bin?
He 2–461	Neb.	N	Y	Y	Y	B[e]	
MWC 342	Neb.	N	Y	Y	Y	B[e]	Symb?/Bin?
MWC 349		Y	N	Y	Y	B[e]	Symb?/Bin?
MWC 645		Y	Y	Y	Y	B[e]	

logue of early type emission lines stars cross-identified against the IRAS point source catalogue (PSC). In turn, these catalogues themselves are selections from millions of stars which combined give us a large original

TABLE 2. Non-supergiant and unclassified B[e] candidates with optical strong Fe II & weak or possible [Fe II] spectral lines

Object	Neb./MC/SF	Isolated	IRAS	NIRE	[O I]	Sp.T.	Remark
New:							
V594 Cas	SF	(Y)	Y	Y	Y	B8:eq	[ ]?
HD 85567		Y	Y	Y	Y	B5vne	[ ]?
V1686 Cyg	SF	(N)	Y	Y	Y	B2/3e	[ ]?
HD 200775	SF	(Y)	Y	Y	Y	Be/3eq	[ ]?
AS 116		Y	Y	Y	Y	B[e]:	
He 3–781		Y	Y	Y	Y	B2V[e]	
V510 Ori	SF	(N)	Y	Y	Y	TT?	[ ]?
HS Ori	SF	(N)	Y	Y	Y	TT?	[ ]?
RW Auri	SF	(N)	Y	Y	Y	G5:V:e=TT?	[ ]?
Confirmed:							
MWC 300		Y	Y	Y	Y	B[e]	In Obs. region
MWC 297	Neb.	Y	Y	Y	Y	O9e	[ ]?
MWC 1080	SF	(Y)	Y	Y	Y	B0eq	
He 3–40		N	N	Y	Y	[ ]	Symb!!
HD 316285	Neb./MC	N	Y	N	Y	B2/3[e]+sh	Symb?/Bin?
He 3–1475		Y	Y	Y	Y	Be	[ ]?
MWC 930		Y	Y	Y	Y	Be	[ ]?
Allen & Swings:							
MWC 137	Neb.	N	Y	Y	Y	B0ep	[ ]?/Symb?/Bin?
AS 321		Y	Y	..	Y	B[e]	
MWC 314	Neb.	Y	Y	..	Y	B[e]	Supergiant?

sample of several 10^3 Be stars of somewhat bright visual magnitudes and with often IRAS point source detections.

We, therefore, have a list of nearly all possible B[e] candidates within the early-type emission lines stars with near- and/or far-IR excess. It is yet not clear and must be studied if B[e] stars without an IR-excess exists, see the discussion in de Winter & Pérez (this proceedings). The only prior selection made in our database is the deletion of B[e] stars that are confirmed LBVs, supergiants, symbiotics and post-AGB stars.

From the selected 322 candidate HAeBe/B[e] stars we possess low-resolution spectra for 191 Southern objects. Statistics from these spectra, show that 51 (27%) present emission lines of Fe II, 66 (35%) of [O I] and 23 (12%) of [Fe II]. From more than 50 early-type EELOs, we find 13 B[e] candidates with strong Fe II and possible or weak [Fe II], and 16 B[e] stars with both Fe II and [Fe II] emissions. We note that nearly all have [O I] emission.

3. Results

A complete list of all new and confirmed B[e] stars and candidates are given in Tables 1 and 2. For those with no available spectra, we have adopted the original description and selection of Allen & Swings (1976), and they were also included. For all these stars, additional observational properties are also listed.

The results of the B[e] stars in Table 1 indicates the unclassified nature of most of these 31 sources; only four might be young, one is probably an LBV, two possible post-AGBs and six symbiotic or binary systems. The other 18 sources remaining are mostly isolated, but seven have a nebular nature, which may suggest that they possible belong to one of the classes described before.

Table 2 lists 19 B[e] candidates from which eight might be young objects, two are probably symbiotic stars or binaries and one is possibly a supergiant. The additional eight sources have no indication of belonging to either of the known homogeneous B[e] groups. We note again that these are all rather isolated sources.

4. Discussion, the classification scheme of B[e] candidates

The classification scheme, presented in Table 3, is very simple. Strictly selected on spectroscopic properties one has a group of B[e] candidates. If other characteristics are known, such as the luminosity, they can be classified, e.g. as LBV or as galactic supergiant B[e] stars (equal as those observed in the MCs, see the review of Zickgraf in this proceedings). From additional properties, we can differentiate whether they are young, part of a symbiotic system or extremely evolved. However, as it can be noticed in literature, and this proceedings, the classification of objects belonging to these groups can be very difficult in some individual cases. Only more and better observations can help to classify them, but until then, they remain as temporarily unclassified B[e] stars. It is, therefore, no coincidence that most of these objects listed in Tables 1 and 2, are only poorly studied!!

Besides this very heterogeneous "group" of B[e] objects, the only new class we find are the B[e] main-sequence objects. These objects are clearly older but they do not show evidences of being in an evolved state. Members of this low-mass BV[e] stellar group are HD 45677 and HD 50138. Considering the enormous amount of objects in our original samples added to the fact that this group contains only two objects with not many candidates left, strongly support the conclusion that no new galactic stellar B[e] class could be defined, when accepting the standard evolutionary scenarios of B

TABLE 3. Schematic presentation of the classification of B[e] stars

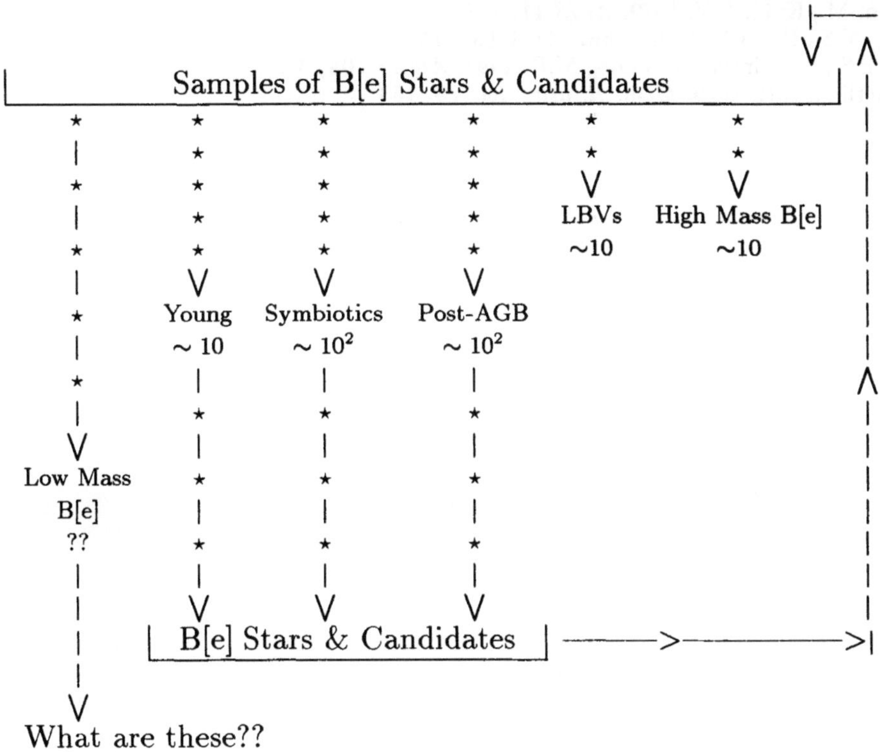

stars. We suggest that BV[e] stars, if not otherwise classified, could be part of binary systems.

Based on our classification scheme we propose to define a new designation for B[e] stars, similarly as done by Lamer et al. (this proceedings). We then separate Herbig B[e] stars, Symbiotic B[e] stars, post-AGB B[e], B[e] LBVs, B[e] supergiants (BI[e]) and B[e] main-sequence (BV[e]) stars. The remaining unclassified objects must be described only with a spectroscopic designation. We propose to use the notation Be[ ] analogous to Bep or Aep, instead of B[e] stars.

Acknowledgements

D. de Winter is supported in part by the Spanish grant DGICYT PB94-0165.

References

Allen, D.A., Swings, J.P. 1976, *A&A* **47,** 293
Cohen, M., Kuhi, L.V. 1979, *ApJS* **41,** 743
Dong, Y.S., Hu, J.Y. 1991, *Chin. A&A* **15,** 275
Thé, P.S., de Winter, D., Pérez, M.R. 1994, *A&AS* **104,** 315
Wackerling, L.R. 1970, *Mem. R. astr. Soc.* **73,** 153

A REVISED CLASSIFICATION SCHEME FOR B[e] STARS

HENNY J.G.L.M. LAMERS
Astronomical Institute, Utrecht University, and SRON Laboratory for Space Research, Princetonplein 5, NL-3584 CC, Utrecht, The Netherlands (hennyl@sron.ruu.nl)

FRANZ-JOSEF ZICKGRAF
Observatoire de Strasbourg, 11 rue de l'Université, F-67000, Strasbourg, France (zickgraf@astro.u-strasbg.fr)

DOLF DE WINTER
Dpto. Física Teórica, C–XI, Facultad de Ciencias, Universidad Autónoma de Madrid, Cantoblanco, E-28049, Madrid, Spain (dolf@astro.up.pt)

LEO HOUZIAUX
Dept d'Astrophysique, Université de Mons-Hainaut, B-7000, Mons, Belgium (leo.houziaux@umh.ac.be)

AND

JANEZ ZOREC
Institute, d'Astrophysique de Paris, C.N.R.S., 98bis bd. Arago, F-75014, Paris, France (zorec@iap.fr)

Abstract.
We review the classification criteria for the B[e] stars. These criteria can be met in different kinds of stars of different evolutionary phases. We propose a new definition of five classes of stars which show the B[e] phenomenon, with names that indicate the evolutionary phase:

(a) B[e] supergiants or "sgB[e] stars"
(b) pre-main sequence B[e]-type stars or "HAeB[e] stars"
(c) compact planetary nebulae B[e]-type stars or "cPNB[e] stars"
(d) symbiotic B[e]-type stars or "SymB[e] stars"
(e) unclassified B[e]-type stars or "unclB[e] stars"

If a star meets the criteria of more than one of the classes (a), (b), (c) and (d) it is automatically assigned to class unclB[e]. [1]

[1] The full paper will be published in Astronomy and Astrophysics.

A. M. Hubert and C. Jaschek (eds.), B[e] Stars, 277–281.

TABLE 1. The classification of B[e] stars

Name	Spectral type	log L_* $[L_\odot]$	B[e] class	Remark
AS 119	B8/9[e]		unclB[e]	
AS 269	B[e]		unclB[e]	SymB[e]?/Bin?
CD-42°11721	B0[e]p		unclB[e]	[FeII] var?
CPD-52°9243	~B3Ia	5.5	sgB[e]	
CW Leo = PK 221+45 1	C?	4.18	cPNB[e]	
GG Car	B0-B2	4.7	sgB[e]	
Hb 12 = PK 111-02 1	?		cPNB[e]	
HD 45677	B2V:[e]		unclB[e]	extreme Be?
HD 50138	B6III-IV[e]		unclB[e]	extreme Be?
HD 51585	B[e]q		unclB[e]	SymB[e]?/Bin?/cPNB[e]?
HD 85567	B5Vne	>2.54	HAeB[e]	
HD 87643	B4II[e]	4.2-4.9	unclB[e]	sgB[e]?/HAeB[e]?
HD 200775	B2/3eq	3.90	HAeB[e]	
HD 326823	B1.5[e]		unclB[e]	
He 2-34 = PK 274+02 1	B[e]		cPNB[e]	
He 2-79	B[e]		unclB[e]	
He 2-90 = PK 305+01 1	?e	3.00	cPNB[e]	
He 2-91	B[e]		unclB[e]	PN?
He 2-139 = PK 326-01 1	B[e]		cPNB[e]	
He 2-446	B:[e]	–	HAeB[e]	
He 2-461	B[e]		unclB[e]	
He 2-1312 = PK 334-07 1	?		cPNB[e]	
He 3-52	B[e]		unclB[e]	
He 3-847	B6/A0:[e]		unclB[e]	
He 3-1138	B8Ia[e]		unclB[e]	LBV?
He 3-1191	B0:[e]		unclB[e]	Post-AGB?
He 3-1475	B[e]	4.34	cPNB[e]	
Hen S12	B0.5	5.3	sgB[e]	
Hen S18		5.5-5.7	sgB[e]	
Hen S22 = HD 34664	B0-B0.5	5.8-6.0	sgB[e]	
Hen S23 = AV 172	B8	4.5	sgB[e]	
Hen S35	B1Iab	5.2	sgB[e]	
Hen S59	B5II	4.0	sgB[e]	
Hen S65 = R50	B2-3	5.7	sgB[e]	
Hen S93	B9Ib	4.7	sgB[e]	
Hen S111 = HDE 269599s	B:	–	sgB[e]	
Hen S134 = HD 38489	B0	5.9	sgB[e]	
Hen S137	B6Ib/II	4.2	sgB[e]	
HK Ori	B/G:[e]p	–	HAeB[e]	

TABLE 2. The classification of B[e] stars (continued)

Name	Spectral type	log L_* $[L_\odot]$	B[e] class	Remarks
M 1-26 = PK 358-00 2	?	4.59	cPNB[e]	
M 2-9 = PK 010+18 2	B[e]	2.82	cPNB[e]	
MWC 17	B[e]		unclB[e]	SymB[e]?/Bin?
MWC 84	B[e]		unclB[e]	cPNB[e]?
MWC 300	B1Ia$^+$	5.7	sgB[e]	
MWC 342	B[e]		unclB[e]	SymB[e]?/Bin?
MWC 349A	B[e]	4.5-5.2	unclB[e]	sgB[e]?/HAeB[e]?
MWC 623	B[e]		unclB[e]	SymB[e]?/Bin?
MWC 645	B[e]		unclB[e]	
MWC 819	B[e]		unclB[e]	
MWC 922	B[e]		unclB[e]	
MWC 939	B[e]		unclB[e]	[FeII]var?
MWC 1080	B0eq	4.64 – 5.33	HAeB[e]	
Mz 3 = PK 331-01 1	B0		cPNB[e]	
OY Gem = HD 51585	B0[e]		cPNB[e]	
PN M 1–92	B0.5IV[e]		unclB[e]	
R 4	B0	5.0	sgB[e]	
R 66 = HDE 268835	B8	5.5	sgB[e]	
R 82 = HDE 269217	B2-3	5.4	sgB[e]	
R 126 = HD 37974	B0.5	6.1	sgB[e]	
R Mon	B0[e]	3.78	HAeB[e]	
SS 73–170	B[e]		unclB[e]	
Th 35–27	B[e]		unclB[e]	
V 380 Ori	B9[e]+sh	2.06	HAeB[e]	
V 594 Cas	B8:eq	4.34	HAeB[e]	
V 704 Cen = PK 311+03 1	B[e]	3.78	cPNB[e]	
V 1016 Cyg	B[e]		unclB[e]	SymB[e]?/cPNB[e]?
V 1686 Cyg	B2/3e	3.56	HAeB[e]	
Vy 2-2 = PK 45-02 1	B[e]		cPNB[e]	

Discussion

Araujo: From the theoretical point of view I did like to remember that we have NO accurate calculations of the radiative force for B-type stars, specially no calculations outside spherical symmetry. As observations are concerned I think we should try to estimate chemical abundances of B[e] supergiants, this would help us to determine their evolutionary phase.
Lamers: One group in Utrecht is calculating the radiative force in a grid of models of different temperatures, densities, etc to understand and explain the bi-stability of Be stars.
Friedjung: Firstly this meeting is neither on protoplanetary nebulae nor on symbiotics. There do exist empirical models particularly for symbiotics. Secondly, you must be careful about making deductions about luminosities of B[e] supergiants. If the disks are optically thick in the continuum, there may be inclination effects on the observed luminosity.
Jaschek: I think I can agree with most of your comments,except the one about "forgetting about details". With the very small number of B[e] stars which have been analysed (8 out of 30?), I think it should be taken with a "grain of salt" - let us collect all possible details and at the next meeting "forget about details".
Oudmaijer: With regard to the CS mass determinations:to fit the SED's with radiative transfer models, a useful tool to try to determine the dust mass and total mass in the CS material.
Cassinelli: Why is it that we see B[e] stars , but not O[e] or A[e] stars?
Oudmaijer: With regard to the A[e] stars: the supergiant IRC +10 420 which changes its spectral type from F8 I in 1973 to A1 now, certainly fulfills the observational criteria: it has strong IR excess, strong Hα, Fe II, [Fe II] and [O I] emissions. It may even be on its way to the B[e] phenomenon.
Zickgraf: In general I believe that the sketch of the model for B[e] supergiants has still evolved; i.e. the evidence for an equatorial disk with dust and a high velocity polar wind still is good. One of the main problems is whether rotation plays a major role in the formation of this structure. The observational determination of rotational velocities is still lacking. With regard to the evolutionary time scales involved, we had made an estimate for the duration of the B[e] phase for the SMC objects. From the member ratio of B[e] supergiants to main sequence objects we estimated a duration of about 10^5 years. That would agree with Langer's main sequence scenario.
Lanz: If I remember correctly, one needs a density larger than 10^{10} part per cm^3 to form CO. A rough order of magnitude estimate would indicate that there is several solar masses in the disk. This estimate must be checked however.

Viotti: Making reference to the history of past colloquia on Be stars, I remind that at that time it turned out the need for a better knowledge of the stellar core. I think that also now what we need for the low luminosity B[e] stars is a deep study of the photospheric stellar spectrum of and to derive the physical properties(Teff, gravity, composition) of the stars.

AUTHOR INDEX

OBJECT INDEX

MIX
Papier aus verantwortungsvollen Quellen
Paper from responsible sources
FSC® C105338

If you have any concerns about our products,
you can contact us on
ProductSafety@springernature.com

In case Publisher is established outside the EU,
the EU authorized representative is:
Springer Nature Customer Service Center GmbH
Europaplatz 3, 69115 Heidelberg, Germany

Printed by Libri Plureos GmbH
in Hamburg, Germany